Fire Safety Aspects
of
Polymeric Materials

VOLUME 2
TEST METHODS, SPECIFICATIONS AND STANDARDS

Report of

The Committee on Fire Safety
Aspects of Polymeric Materials

NATIONAL MATERIALS ADVISORY BOARD
Commission On Sociotechnical Systems
National Research Council

Publication NMAB 318-2
National Academy of Sciences
Washington, D.C.
1979

VOLUME 2 – TEST METHODS, SPECIFICATIONS, AND STANDARDS

Fire Safety Aspects of Polymeric Materials

265 Post Road West, Westport, CT 06880

Printed in U.S.A.
Library of Congress Card No. 77-79218
I.S.B.N. 0-87762-223-4

NOTICE

This study by the National Materials Advisory Board was conducted under Contract No. 4-35856 with the National Bureau of Standards.

Printed in the United States of America.

FOREWORD

This volume is one of a series of reports on the fire safety aspects of polymeric materials. The work reported here represents the results of the first in-depth study of this important subject. The investigation was carried out by a committee of distinguished polymer and fire technology scholars appointed by the National Academy of Sciences and operating under the aegis of the National Materials Advisory Board, a unit of the Commission on Sociotechnical Systems of the National Research Council.

Polymers are a large class of materials, most new members of which are man-made. While their versatility is demonstrated daily by their rapidly burgeoning use, there is still much that is not known or not widely understood about their properties. In particular, the burning characteristics of polymers are only now being fully appreciated and the present study is a landmark in the understanding of the fire safety of these ubiquitous materials.

In the first volumes of this series the committee has identified the limits of man's knowledge of the combustibility of the growing number of polymeric materials used commercially, the nature of the by-products of that combustion, and how fire behavior in these systems may be measured and predicted. The later volumes deal with the specific applications of polymeric materials, and in all cases the committee has put forth useful recommendations as to the direction of future actions to make the use of these materials safer for society.

Harvey Brooks, Chairman
Commission on Sociotechnical Systems

ABSTRACT

This is the second volume in a series. The fire safety aspects of polymers are examined with primary emphasis on human survival. Other volumes in the series deal with materials: state of the art; fire dynamics and scenarios; aircraft (civil and military); and applications to buildings, vehicles, ships, mines and bunkers. A summary volume (Vol. 5) has been added to the series to pull together the disciplinary material of the first four volumes.

This report summarizes the state of the art on test methods which can be employed for evaluating various aspects of flammability behavior in polymeric materials. Separate chapters in the report are concerned with a broad overview of problems associated with the development of tests and standards (Chapter 3), with brief summary of the most important U.S. test methods (Chapter 4), and with a critical discussion of these (Chapter 5) in which status, limitations and problems of currently available test methods are reviewed. The concepts of fire dynamics, and their importance in the future development of meaningful test methods are discussed under a separate heading (Chapter 6).

In this report, test methodology is reviewed with cursory reference to the end uses where a particular test may be employed or applicable. A more detailed discussion of test methods used for specific products (e.g., building materials, coatings, fabrics, etc.) is provided in the appropriate end-use volumes.

Conclusions and recommendations are included at the end of Chapters 5, 6, and 7, and also highlighted in Chapter 2.

VOLUMES OF THIS SERIES

Volume 1	Materials: State of the Art
Volume 2	Test Methods, Specifications and Standards
Volume 3	Smoke and Toxicity (Combustion Toxicology of Polymers)
Volume 4	Fire Dynamics and Scenarios
Volume 5	Elements of Polymer Fire Safety and Guide to the Designer
Volume 6	Aircraft (Civil and Military
Volume 7	Buildings
Volume 8	Land Transportation Vehicles
Volume 9	Ships
Volume 10	Mines and Bunkers

PREFACE

The National Materials Advisory Board (NMAB) of the Commission on Sociotechnical Systems, National Research Council, National Academy of Sciences-National Academy of Engineering, was asked by the Department of Defense, Office of Research and Engineering, and the National Aeronautics and Space Administration to "initiate a broad survey of fire-suppressant polymeric materials for use in aeronautical and space vehicles, to identify needs and opportunities, assess state of the art in fire-retardant polymers (including available materials, production, costs, data requirements, methods of test and toxicity problems), and describe a comprehensive program of research and development needed to update the technology and accelerate application where advantages will accrue in performance and economy."

In accordance with its usual practice, the NMAB convened representatives of the requesting agencies and other agencies known to be working in the field to determine how, in the national interest, the project might best be undertaken. It was quickly learned that duplication of interest existed. At the request of the other agencies the project scope was broadened, subject to satisfying the needs of the original requestors.

The total list of sponsors of this study now comprises Department of Agriculture, Department of Commerce (National Bureau of Standards), Department of Interior (Division of Mine Safety), Department of Housing and Urban Development, Department of Health, Education and Welfare (National Institute for Occupational Safety and Health), Department of Transportation (Federal Aviation Administration, Coast Guard), Department of Energy, Consumer Product Safety Commission, Environmental Protection Agency, and Postal Service, as well as the original sponsors.

The committee was originally constituted on November 30, 1972 and expanded to its present status on July 26, 1973. The new scope was established after presentation of reports by liaison representatives covering needs, views of problem areas, current activities, future plans, and relevant resource materials. Tutorial presentations were made both at meetings held in the Academy and during site visits, when the committee or its panels met with experts and organizations concerned with fire safety aspects of polymeric materials. These site visits (upwards of a dozen) were an important feature of the committee's search for authentic information. Additional inputs on foreign fire technology were supplied by the U.S. Army Foreign Science and Technology Center and NMAB Staff.

This study in its various aspects is addressed to those who formulate policy and allocate resources. A sufficient data base and bibliography has been supplied to indicate the breadth of this study.

ACKNOWLEDGMENTS

This report was drafted by the committee's panel on "Test Methods, Specifications and Standards" and by a number of government liaison representatives. It was then reviewed by the entire committee, which has responsibility for all the conclusions and recommendations included in the report. The following alphabetical listing of panel members and liaison representatives is intended to include all those who contributed to the actual preparation of the report: Raymond Friedman, Carlos Hilado, Clayton Huggett, Jurgen Kruse, Irving Litant, Richard Magee, Donald Moore, Daniel Pratt, Jack Ross, Giuliana Tesoro and Arnold Weintraub. Overall coordination of the draft of this volume was furnished by Professors Magee and Tesoro.

Many other individuals have contributed significantly through discussion and advice, and their assistance is gratefully recognized. A listing will not be attempted (since it would be impossible to do justice to all contributors).

Several organizations have contributed by providing demonstrations of ongoing development work on test methods and discussions/presentations on the relationships of test results to hazard in fire situations. Special thanks are due to the Factory Mutual Research Corporation (Paul Cotton), the Underwriters Laboratory (Mr. Castino), the U.S. Army Quartermaster Laboratories, and University of California, Berkeley (Prof. Williamson).

I acknowledge with gratitude the assistance in this project of Dr. Robert S. Shane (NMAB Staff Scientists) and Miss Carolyn A. Tuchis, our able secretary.

Dr. Herman F. Mark, Chairman

NATIONAL MATERIALS ADVISORY BOARD
COMMITTEE ON FIRE SAFETY ASPECTS OF POLYMERIC MATERIALS

Chairman

HERMAN F. MARK, Emeritus Dean, Polytechnic Institute of New York, Brooklyn.

Members

ERIC BAER, Head, Division of Macromolecular Science, Case-Western Reserve University, Cleveland, Ohio

ALFRED E. BROWN, Celanese Corporation, New York.

JOHN M. BUTLER, Manager, Organic Polymer Synthesis, Monsanto Research Corp., Dayton, Ohio.

RAYMOND FRIEDMAN, Vice President and Scientific Director, Factory Mutual Research Corp., Norwood, Massachusetts.

ALFRED R. GILBERT, Manager, Organic and Polymer Chemistry Branch, Chemistry Laboratory, Corporate R&D, General Electric Company Schenectady, New York.

CARLOS J. HILADO, Research Professor, Institute of Chemical Biology, University of San Francisco, California.

RAYMOND R. HINDERSINN, Section Manager, Research Center, Hooker Chemical Co., Niagara Falls, New York.

WILLIAM C. HUSHING, Vice President, Kensington Management Consultants, Bath, Maine.

FRANK E. KARASZ, Professor and Co-Director, Materials Research Laboratory, Polymer Science & Engineering Dept., University of Massachusetts, Amherst.

TED A. LOOMIS, Professor & Chairman, Department of Pharmacology, School of Medicine, University of Washington, Seattle.

RICHARD S. MAGEE, Associate Professor, Department of Mechanical Engineering, Stevens Institute of Technology, Hoboken, New Jersey.

ELI M. PEARCE, Head, Department of Chemistry, Professor of Polymer Chemistry and Chemical Engineering, Polytechnic Institute of New York, Brooklyn, New York.

CHARLES F. REINHARDT, Director, Haskell Laboratory for Toxicology and Industrial Medicine, E. I. duPont Nemours and Co., Newark, Delaware.

ARNOLD ROSENTHAL, Celanese Research Company, Summit, New Jersey.

DONALD G. SMILLIE, Chief Structures Development Engineer, McDonnell Douglas Corp., Long Beach, California.

GIULIANA C. TESORO, Adjunct Professor, Fibers and Polymers Laboratories, Massachusetts Institute of Technology, Cambridge, Massachusetts

NICHOLAS W. TSCHOEGL, Professor Department of Chemical Engineering, Division of Chemistry and Chemical Engineering, California Institute of Technology, Pasadena California.

ROBERT B. WILLIAMSON, Associate Professor of Engineering Science, Dept. of Civil Engineering, University of California, Berkeley, California.

Liaison Representatives

JEROME PERSH, Staff Specialist for Materials and Structures, ODUSDRE(ET), Department of Defense, Washington, D.C.

ERNEST B. ALTERKRUSE, Chief, Department of Clinics, Moncrief Army Hospital, Fort Jackson, Columbia, South Carolina.

ALLAN J. McQUADE, U.S. Army Natick Laboratories, Natick, Massachusetts.

GEORGE R. THOMAS, Department of the Army, Army Materials and Mechanics Research Center, Watertown, Massachusetts.

HERBERT C. LAMB, Naval Facilities Engineering Command, Alexandria, Virginia.

DANIEL PRATT, Naval Ships Engineering Center, Hyattsville, Maryland.

GEORGE SORKIN, Naval Sea Systems Command, Washington, D.C.

JACK ROSS, Wright-Patterson Air Force Base, Dayton, Ohio.

BERNARD ACHHAMMER, National Aeronautics and Space Administration, Advanced Research and Technology Division, Washington, D.C.

JOHN A. PARKER, National Aeronautics and Space Administration, Ames Research Center, Moffett Field, California.

ARNOLD WEINTRAUB, Department of Energy, Washington, D.C.

DAVID FORSHEY, Department of the Interior, Bureau of Mines, Washington, D.C.

JOSEPH E. CLARK, National Fire Prevention Control Agency, Department of Commerce, Washington, D.C.

CLAYTON HUGGETT, Center for Fire Research, National Bureau of Standards, Washington, D.C.

LESLIE H. BREDEN, Fire Safety Engineering Division, Center for Fire Research, National Bureau of Standards, Washington, D.C.

IRVING LITANT, Department of Transportation (DoT), Systems Center, Kendall Square, Cambridge, Massachusetts.

GEORGE BATES, JR., Department of Transportation, Federal Aviation Administration, Washington, D. C.

RALPH RUSSELL, Aircraft Division, DoT, Federal Aviation Administration, National Aviation Facilities Experimental Station, Atlantic City, New Jersey.

PAUL W. SMITH, Aviation Toxicology Laboratory, Civil Aeromedical Institute, Oklahoma City, Oklahoma.

ROBERT C. McGUIRE, DoT, Federal Aviation Administration, Washington, D. C.

THOMAS G. HOREFF, DoT, Federal Aviation Administration, Washington, D. C.

DANIEL F. SHEEHAN, DoT, U. S. Coast Guard, Washington, D. C.

WILLIAM J. WERNER, Department of Housing and Urban Development, Washington, D. C.

DONALD L. MOORE, Department of Housing and Urban Development, Washington, D. C.

IRVING GRUNTFEST, Environmental Protection Agency, Washington, D. C.

NELSON GETCHELL, U. S. Dept. of Agriculture, Beltsville, Maryland.

WILSON A. REEVES, Cotton Finishes Laboratory, U. S. Dept. of Agriculture, New Orleans, Louisiana.

HERBERT W. EICHNER, Forest Service, U. S. Dept. of Agriculture, Forest Products Laboratory, Madison, Wisconsin.

RICHARD E. WIBERG, National Institute for Occupational Safety and Health, Rockville, Maryland.

JAMES RYAN, Consumer Product Safety Commission, Bethesda, Maryland.

JURGEN KRUSE, Materials Research Div., Office of Postal Science and Technology, Rockville, Maryland.

Technical Advisors

IRVING N. EINHORN, Flammability Research Center, Division of Materials Science and Engineering, Flammability Research Center, College of Engineering, University of Utah, Salt Lake City.

LAWRENCE M. ENGLEMAN, Washington Metropolitan Area Transit Authority, Washington, D. C.

MAUREE W. AYTON, Science Information Analyst, Congressional Research Service, Library of Congress, Washington, D. C.

National Materials Advisory Board Staff

ROBERT S. SHANE, Consultant

Technical Editor

ISRAEL KATZ, Professor of Engineering Technology, Northeastern University, Boston, Mass.

CONTENTS

CHAPTER 1

INTRODUCTION

1.1 Scope and Methodology of the Study

Consistent with its charge, the committee directed its attention to the behavior of polymeric materials in fire situations with special emphasis on human safety considerations. Excluded from those considerations, however, were firefighting, therapy after fire-caused injury and mechanical aspects of design not related to fire safety.

The work of the committee included: (1) a survey of the state of pertinent knowledge, (2) identification of gaps in that knowledge, (3) identification of work in progress, (4) evaluation of work as it relates to identified gaps, (5) development of conclusions, (6) formulation of recommendations for action by interested public and private agencies, and (7) estimation, where appropriate, of the benefits that might accrue through implementation of the recommendations. Within this framework, fundamental disciplinary areas were addressed individually; end-uses were considered when the possibility of fire was a design consideration and the end-uses were of concern to the sponsors of the study.

Attention was given to natural and synthetic polymeric materials; primarily in terms of their fire safety performance as determined by their composition, structure, reaction to processing, and geometry (i.e., film, foam, fiber, etc.); but special aspects relating to their incorporation into an end-use component or structure also were included. Relevant test methods, specifications, and standards were considered; regulations, however, were dealt with only in relation to end-uses.

The products of combustion, including smoke and toxic substances, were considered in terms of their effects on human safety; morbidity and mortality were treated only as a function of the materials found among the products of combustion. The question of potential exposure to fire-retardant polymers, including skin contact, in situations not including pyrolysis and combustion were addressed as deemed appropriate by the committee in relation to various end uses.

In an effort to clarify the understanding of phenomena accompanying fire, attention was directed to the mechanics of mass and energy transfer (fire dynamics). The opportunity to develop one or more scenarios to guide thinking was provided; however, as noted above, firefighting was not considered. To assist those who might use natural or synthetic polymers in components or structures, design principles and criteria were also considered.

In organizing its work, the committee concluded that its analysis of the fire safety of polymeric materials should consider the materials themselves, the fire dynamics and the societal systems affected. This decision led to the development of a report structure that provides for separate treatment of the fundamental-disciplinary aspects of the problem (the data base) and various aspects of product end-use.

Accordingly, the committee presents its findings in five disciplinary and five end-use reports as follows:

Volume 1	Materials: State of the Art
Volume 2	Test Methods, Specifications and Standards
Volume 3	Smoke and Toxicity (Combustion Toxicology of Polymers)
Volume 4	Fire Dynamics and Fire Scenarios
Volume 5	Elements of Polymer Fire Safety and Guide to the Designer
Volume 6	Aircraft (Civil and Military)
Volume 7	Buildings
Volume 8	Land Transportation Vehicles
Volume 9	Ships
Volume 10	Mines and Bunkers

Some of the polymer applications and characteristics are in the classified literature; the members of the committee with security clearances believed that this information could best be handled by special meetings and addendum reports to be prepared after the basic report volumes were completed. Thus, the output of the committee is freely available to the public. Considering the breadth of the fire safety problem, it is believed that exclusion of classified information at this time will not materially affect the committee's conclusions.

Many statements about the fire safety aspects of polymeric materials appear in each of the committee's reports, but members of the committee emphasize that such statements, including judgmental comments regarding fire safety aspects of materials in particular end-uses, apply only to the specific situations they describe. The suitability of a material, from a fire safety point of view in a given application, depends on many factors, e.g., ease of access, ease of egress, proximity of ignition hazard, proximity of other combustible materials, thermal flux and duration of ignition source, ambient oxygen partial pressure, and fire and smoke detection and suppression systems in place. Therefore, statements in this volume (as in the others) must not be taken out of context and applied to the use of identical materials in other situations.

Members of the committee are and have been involved with materials research, development and applications as well as system design and evaluation, whereas the sponsors' liaison representatives deal with materials research and development, regulation, procurement, operations and analysis. In these circumstances, the fire safety performance of each material considered was subjected to a broad spectrum of expertise. Full and extensive communication over the lengthy period of the committee's deliberations provided an unusual base for augmentation of its expertise and rounding of knowledge.

1.2 Scope and Limitations of This Report

In this volume, the "state of the art" is reviewed in a general way. Technical problems of fire testing in polymeric materials, and some of the reasons for lack of correlation between results of laboratory tests and behavior of materials under use conditions are discussed in Chapter 3. Most well-documented test methods currently used for commercial products are outlined in Chapter 4 and critically discussed in Chapter 5. Theoretical knowledge on which future test method development might be based is presented in Chapter 6. The agencies and organizations that are currently responsible for the formulation and enforcement of fire safety specifications and standards are indicated in Chapter 7.

Throughout this volume, the subject is discussed broadly to present an overall view of the problem. Detailed reviews of test methods specifically designed for or applicable to specific end-use requirements (e.g., aircraft cabins, building materials, etc.) are presented separately in the appropriate end-use volumes.

Primarily, the material reviewed covers U. S. technology. It is not intended to ignore advances in foreign countries (in fact, salient test methods developed abroad are included), but it is necessary to recognize that the state of the art on technology of test methods cannot be reviewed without concomitant consideration of the societal factors and legislative processes that motivate and determine the evolution of fire safety standards for polymeric materials.

In attempting a critical evaluation of the state of the art, rather than the preparation of a mere catalog of existing methodology, the committee has (inevitably) exercised judgment in the selection of technical information available in the literature and from other sources. In doing so, every effort has been made to avoid bias, and to reach an objective, unified, coherent assessment of the state of the art on this critically important aspect of the fire safety problem.

1.3 General Considerations

As concern for fire safety increases in our society, requirements for the response of polymeric materials to heat and fire must be defined in the context of specific end use, and of potential exposure to a wide range of fire conditions. This task poses formidable problems. Even with the best scientific, technological, sociological and political skills, approaches to these problems are barely discernible at this time. Progress would be enhanced if the flammability of or fire hazard posed by a given polymeric material were an intrinsic property that could be measured in definitive terms. In fact, flammability of polymeric materials and fire hazard posed by specific products of systems depend greatly on many factors in addition to composition. Geometry, orientation, ventilation, proximity of other materials, etc., can have overriding effects on the response of a material in a fire.

Also, progress might be accelerated if measurements of relative flammability made for a given material in the laboratory could be interpreted and analyzed to provide accurate indications of the material's behavior in a major fire. This is not the case, since, to date, our understanding of the physics and chemistry of fires and of fire phenomena has not been sufficient to provide a basis for definitive simulations in the laboratory.

The subjects of testing methodology, validity and significance of test results, interpretation of laboratory results, and extrapolation to end-use conditions are of critical importance in the overall consideration of the fire safety of materials. Technical efforts of many groups have produced important advances, and the rationale for the selection of specific test methods is being reviewed and revised on a continuing basis. Nevertheless, incomplete scientific knowledge and limitations in dealing with many problems of fire testing are reflected in a continuing pragmatism that has led to the development of scores of test methods, each designed to simulate use conditions a little better, or to measure differences between materials or trends more precisely, or to isolate specific parameters of flammability behavior, or to provide experimental verification of idealized theoretical models.

CHAPTER 2

SUMMARY OF CONCLUSIONS AND RECOMMENDATIONS

2.1 Summary

The behavior of a polymeric material in a fire environment and the hazard posed by a given product depend on a number of factors (in addition to chemical composition). Geometry of the system, ventilation, proximity of other materials, etc., determine to a large extent the response of a given material in a fire. The flammability of, or the fire hazard posed by a given polymeric material is not an intrinsic property that can be measured in definitive terms by a simple test.

Early fire tests were developed primarily for naturally occurring polymers, and with very limited understanding of fire science. While knowledge has improved, material technology has progressed at an even more rapid rate, and demands for more meaningful fire tests have outstripped the gains in knowledge.

The objective of all fire safety activities is to reduce loss from unwanted fires. To work towards this objective, the subjects of testing methodology, validity and significance of test results, interpretation of laboratory results, and extrapolation to end-use conditions must be considered of critical importance.

2.2 Conclusions and Recommendations

(This is a partial compilation of the more important conclusions and recommendations. The committee's complete conclusions and recommendations will be found at the end of chapter 5, 6, and 7 as noted).

Conclusion: For a long time, test methods have been developed with limited understanding of the phenomena of fire growth. Knowledge has advanced considerably in the last decade, and the development of more meaningful test methods based on fire dynamics cannot be attempted. *Recommendation:* Require that tests and standards demonstrate a rational relationship to the hazards they are designed to control.

Conclusions: Many test methods and criteria for fire hazard used today were developed through experience with cellulosic materials. These tests are not necessarily adequate for the valuation of fire hazard of synthetic polymers. No single test is adequate to completely evaluate the fire hazard of a particular material. The complexity of *ignition* phenomena makes it impossible to devise a single test for the evaluation of ease of ignition that is applicable to different materials and end-use situations. *Recommendations:* In the development of tests for ignitability, simulate well-defined ignition sources and end-use conditions.

Conclusion: The rate of *flame spread* is strongly dependent on many physical and geometrical parameters. Thus, tests that presume to establish relative flame spread characteristics of polymeric materials in a definitive manner may be valid only for the particular conditions of the test and may give misleading results if extrapolated to other conditions. *Recommendations:* Continue work on tests for fire hazard of interior finish materials; the currently used ASTM E-84 test must be improved or replaced. Correlate *surface flammability* test ratings to full-scale fire hazard.

Conclusion: The rate at which *heat is released during burning is an important criterion for evaluating fire hazard from a particular material. Therefore, tests measuring rate of heat release can be of great significance. Recommendation:* Continue the current development of tests for *rate of heat release.*

Conclusions: The *oxygen index* test provides a laboratory technique for the evaluation and guidance of the development of new materials; however, since it does not model energy feedback realistically, it is not sufficient for characterizing fire hazard or standards or regulations. There are no standardized test methods for determining relative *ease of extinguishment* for different polymeric materials. In view of the many factors influencing extinguishment, there seems to be little chance that a small-scale test method could be developed to adequately determine minimum water application rate for a given polymeric material. Tests for *smoke evolution* are important, since the presence of smoke is an important parameter of fire hazard. Available tests for smoke evolution yield results that depend greatly on the conditions of testing (e.g., ventilation, heat flux level, sample orientation). *Recommendation:* Place high priority on the development of more meaningful tests for *smoke evolution.*

Conclusion: Fire safety toxicity standards for materials have largely ignored the complexity of the problem. No accepted toxicity tests are currently available. *Recommendations:* Mandate the use of laboratory animals during a first-tier screening assessment of combustion product toxicology. Incorporate both a behavioral end-point and a mortality count in an appropriate toxicity screening test.

Conclusion: Currently available tests for fire endurance are generally satisfactory. *Recommendation:* Run the ASTM E-119 fire endurance test with a positive pressure on at least the upper two thirds of wall specimens so as to improve the test when there are openings in the wall.

Conclusion: Large-scale tests are necessary as acceptance tests for polymeric materials and products and to establish the validity of various small-scale tests. *Recommendations:* Reflect the nature and type of fire scenario to be protected against when designing large-scale room tests.

Conclusion: The complex interaction of fundamental processes in fire dynamics, coupled with a relatively small funding effort, has led to rather slow progress in quantifying fire behavior. *Recommendation:* An increased program of sufficient magnitude and stability, funded by the government, in the specific area of fire dynamics must be established. This program should be conducted primarily in academic, non-profit, and governmental research organizations but the projects should be closely monitored by advisory boards including broad representation by manufacturers, users, and members of standards-setting organizations.

Conclusion: The current level of understanding of fire dynamics is not sufficiently developed to permit thorough scientific analysis of most fire test methods. The behavior of a given material in a fire depends not only on the properties of the fuel but also on the fire environment to which it is exposed. Consequently, if test methods are to be meaningful, they must simulate the critical fire dynamic conditions. An understanding of fire dynamics is essential if the critical conditions are to be identified. Consequently, fire dynamics can be extremely useful in test method selection and development. Valid modeling procedures based on evolving fire dynamic principles offer promise for reduction in the dependence on, and the cost of, large-scale fire tests. *Recommendation:* Fire dynamics expertise should be employed when developing new test methods and for validation and improvement of existing test methods.

Conclusion: The major importance of test methods is that they become a part of the specifications, standards, and codes that define the performance of materials. *Recommendation:* Test methods incorporated in regulations must be carefully investigated to establish their validity and limitations.

Conclusion: Only the United States among developed nations has widely fragmented systems of fire safety standards and codes. This contributes to losses of life and property that might be avoided. *Recommendation:* Attack the problem of fragmentation of fire codes by judiciously evaluating the modes used in other developed countries and adopting their practices as seems desirable in the United States.

Conclusion: Codes and standards are only of value to the extent that they are enforced. *Recommendation:* Enforcement of fire codes and mandatory standards should be uniform and rigorous.

Conclusion: Recognizing that all contributors to the fire protection system should be stimulated to maximum activity in the pursuit of fire safety, incentives should be sought to further these activities. *Recommendation:* Search out economic incentives such as favorable insurance rates, favorable tax treatment, or other societal incentives that favor improved fire safety.

CHAPTER 3

OVERVIEW OF TEST METHODS, STANDARDS, SPECIFICATIONS, AND CODES

3.1 Introduction

In the United States the choice of materials relative to fire safety is governed by a complex hierarchy of test methods, standards, specifications, codes and related regulations (Lyons 1975; National Bureau of Standards, 1975). The complexity of this system and the number of agencies involved present serious difficulties, both for the manufacturers of polymeric materials and products and for the users of such products.

The objective of all fire safety activities is to reduce loss from unwanted fires. The precise definition of loss is itself a difficult and controversial problem, but for our present purpose we will focus attention on two major components, human death and injury and the destruction of material property.

Fire safety activities fall into five general categories:

(a) the unorganized efforts of individuals, based on limited knowledge of the problem and concerned with the safety of their own property and the personal safety of their family and associates,

(b) the voluntary efforts of associations concerned with the development of fire test methods and recommended safety practices and the dissemination of authoritative fire safety information,

(c) the economically motivated efforts of industrial organizations, including the insurance industry, concerned primarily with minimizing property loss,

(d) the statutory efforts of government regulatory agencies to reduce the frequency and severity of fires through mandatory codes and standards,

(e) the organized efforts of the Fire Services directed primarily toward the prevention, control, and extinguishment of fires.

Test methods, standards, specifications, and codes have their origin and find their applications principally in activities (b) and (d) above.

Test Methods are procedures for measuring a property or behavioral characteristic of a material, product, or assembly as an aid to predicting its performance in application. Most test methods were designed originally to guide the development or acceptance of a material, product or system. If a method proves useful, it may be described in publications and adopted by others with similar needs. If it is recognized as a generally useful procedure, it may be submitted to a voluntary standards writing and promulgating body such as the American Society for Testing and Materials (ASTM), the National Fire Protection Association (NFPA), a technical society, or trade association for possible adoption. After further consideration, usually including inter-laboratory trials to establish operability and reproducibility, it may be accepted and documented as a standard test method.

The term *"Standards"* is applied to a variety of rules and procedures designed to provide an orderly approach to problems of fire safety. Standards may include definitions and terminology, methods of measurement, performance requirements, and rules for the protection of persons and property. The majority of the standards in use in the United States are Voluntary Standards, developed by legitimizing organizations frequently through a consensus process involving producers, users, and general interest groups (National Bureau of Standards, 1970). Such voluntary standards may become mandatory standards if adopted by a government agency having jurisdiction in the area. Mandatory standards may also be promulgated by government agencies having specific statutory responsibility. For example, the Consumer Product Safety Commission issues standards covering the safety of consumer products while the Department of Transportation establishes regulations governing the safety of vehicles. Such standards may specify a level of performance suitable to the proposed application, a test method to measure performance, and sampling plans, record keeping requirements and other appurtenances necessary to demonstrate compliance with the standard.

Specifications find their principal use in commercial practice where they are used to define the properties of a material, product or structure being procured. They establish the level of performance which the items must meet and the test method by which the performance is to be measured. When incorporated into a purchase contract, the specification becomes part of a legal document enforceable in the courts. The citation of standard test methods in specifications is a great advantage since it obviates lengthy and perhaps ambiguous description of test procedures. Thus, to say that a flame spread classification is to be determined by ASTM Method E 84 is equivalent to including nine pages of descriptive material from the ASTM Book of Standards.

Codes are compilations of standards and recommended practices designed to assure a desired level of safety in a building or building sub-system. Examples are the NFPA Life Safety Code, the National Electric Code, and the four model Building Codes. These are usually voluntary codes developed by the consensus process, but they become mandatory within the designated area of jurisdiction when adopted by a local, state, or federal government agency. Unlike Canada and several other countries, the United States does not have a national building code which can be used to govern fire safety as well as other aspects of building construction on a nationwide basis. Instead, these requirements are covered by some 20,000 local codes. While most of these codes are based on one or another of the model codes and possesss a considerable degree of similarity, local differences pose a serious problem to architects, designers, material suppliers and builders.

Guidelines may be issued by a particular agency or institution in expectation of more precise specifications and/or standards to follow. They generally represent compendia and/or lists of requirements to be met by products, processes, or systems until such time as regulations are issued. Guidelines cannot be enforced, but they are an important factor in the procurement of materials or systems which are not covered by regulations.

3.2 Present Status of Fire Test Methods Applicable to Polymeric Materials

The field of fire hazard testing of materials is in a state of ferment. The relevance of old, well established test methods is being questioned and new concepts of performance testing are being discussed. This situation is due largely to two factors; an influx of new materials and products, principally based on synthetic polymers, into the environment and an increased emphasis on life safety in fire safety regulations.

Up to about 10 years ago, the emphasis of fire safety regulations, fire test methods, and fire research was on the protection of property. The combustible components of most structures were of natural origin and their fire performance properties were intuitively recognized through long familiarity. A few well established tests served to characterize the behavior of new applications and new construction techniques.

With the recognition that the human cost of fires, in terms of deaths and injuries, might equal or exceed the direct property loss, and with increased emphasis on public safety, the main thrust of fire safety development has shifted from the protection of structures and property to the protection of building interiors and their occupants. This thrust has focused attention on the fire safety properties of a new class of products, the multitude of consumer products, materials and furnishings that create the environments of modern structure interiors. These materials and products provide numerous opportunities for ignition, rapid fire growth, and the spread of smoke and toxic gases. They pose a threat to human safety out of proportion to their contribution to structural damage and property loss in most fires.

Many new products are made of polymeric materials. Moreover, a great variety of new materials whose widely differing (from traditional materials) fire performance properties are not known or not widely recognized have been introduced. Fire performance tests designed to evaluate the structural fire performance of traditional materials were inappropriate for these new materials and applications. As a result, there has been a proliferation of new small scale test methods, most of which originated as ad hoc tests in development laboratories and were designed to permit the convenient comparison of experimental material formulations. Their relevance to the performance of products under use conditions was seldom a factor in their development. A few of these tests have achieved recognition as standard test methods by virtue of their endorsement by ASTM, NFPA, Underwriters Laboratories and other organizations.

The action of the Federal Trade Commission against the cellular plastics industry (Federal Trade Commission 1975) triggered a massive reexamination of the relevance of existing fire test methods to the prediction of the performance of materials, products, and systems in fires. While the FTC action provided the impetus, recognition of the problem had been building up in the fire research community for a number of years.

The proposed FTC complaint alleged: "The aforesaid ASTM test standards are neither reliable nor accurate tests for determining, evaluating, predicting or describing the burning characteristics of plastics products under actual fire conditions."

The crux of the complaint is found in the last three words. What are "actual fire conditions"? To consider this question ASTM established Committee E-39 on Fire Hazard Standards (later merged with Committee E-5). Similar committees have been established in several other industrial countries, and the philosophy of fire hazard testing has become a popular subject for discussion (Malhotra, 1975).

A basic shortcoming of most small scale tests can be illustrated by reference to a typical test method such as ASTM D 1692. Here a small sample of the test material in a horizontal orientation is ignited at one end and the progress of burning is observed. In the combustion of a solid fuel, energy feedback from the high temperature gaseous combustion zone pyrolyzes the fuel surface to provide a continuing supply of gaseous fuel to the flame. The rate of burning is directly related to the magnitude of this energy feedback. In the ASTM D 1692 test configuration, most of the energy of combustion is dissipated in the rising convective plume and through radiation to the cool surroundings. In a real fire, on the other hand, energy exchange between adjacent fuel surfaces and radiation from the heated surroundings greatly increases the energy feedback and the intensity of combustion. Experiments at Factory Mutual Research Corporation have shown that the burning rate of a wood crib in a well ventilated compartment may be nearly twice as great as that of a similar crib burning in the open because of increased energy feedback (Friedman, 1975). In the same laboratory it was found that radiation provided approximately 85% of the energy flux to the burning surface at the top of a vertical 12 ft. plastic wall panel, resulting in a threefold increase in burning rate over that at the bottom of the panel (Orloff et al. 1976). Obviously, any test which pretends to simulate "actual fire conditions" must simulate the energy environment of a real fire.

3.3 Development of Tests and Standards

In the past, test method development and the formulation of codes and standards has proceeded in a largely unstructured and sometimes haphazard fashion. Test methods frequently evolved from the ad hoc tests of the development laboratory and standards were developed intuitively in response to a perceived need. In more recent years the process has begun to become more formalized, and new tests and standards are expected to demonstrate a rational relationship to the hazards they are designed to control. This change is related to two factors. First, the increase in government regulatory activity has made these regulations subject to challenge in the courts, requiring a factual basis for regulation as opposed to the intuitive approach that has served in the past. Second, an increased awareness of the societal costs of regulation is reflected in demands for a high level of cost effectiveness.

The rational development of an effective fire safety standard is a four step process:

1. Identification of Hazard
2. Quantification of Hazard
3. Development of Test Method
4. Development of Performance Standard

3.3.1 Identification of Hazard

The order of magnitude of the fire loss in this country is well known and this alone is sufficient to justify a major effort to reduce the loss. However, the diversity of fire problems requires a much more detailed knowledge of where and how fires occur before effective remedial action can be taken. This information can be supplied by statistical surveys of fire incidents.

A major source of fire incident data is the FIDO (Fire Incident Data Organization) file maintained by the NFPA. This computerized data base contains information on approximately 30,000 fire-related incidents in the period 1971-1975. The data are obtained from the fire services and are primarily from fires causing death, injury or major property loss. Consequently, the FIDO file does not present a total profile of fire experience. For example, deaths and injuries due solely to apparel fires are probably underrepresented because the fire services are not always called to such fires. Small fires which have the potential of growing to major proportions if not quickly controlled are also excluded. Nevertheless, this is the largest source of U.S. data on major fires and is invaluable in the identification of hazard area and the establishment of priorities (National Fire Protection Association, 1975; Clarke and Ottoson, 1976).

In an effort to establish a more statistically valid measure of fire experience, the National Bureau of Standards (NBS) and the Consumer Product Safety Commission (CPSC) jointly sponsored a National Household Fire Survey conducted by the Bureau of the Census (National Fire Prevention Control Administration, 1975). A statistically selected sample of 33,000 households were interviewed and a total of 2,463 fire incidents were reported to have occurred during the preceding year. Follow-up interviews provided additional details on these incidents. While business establishments and public buildings were excluded from the survey, the results give a useful estimate of the magnitude of the household fire problem.

This effort is being continued by the recently established National Fire Data Center of the National Fire Prevention and Control Administration (NFPCA) (Public Law 93-498, 1974). The National Fire Data Center is charged with the responsibility to gather and analyze —

(1) information on the frequency, causes, spread, and extinguishment of fires;
(2) information on the number of injuries and deaths resulting from fires, including the maximum available information on the specific causes and nature of such injuries and deaths, and information on property losses;
(3) information on the occupational hazards faced by fire-fighters, including the causes of deaths and injuries arising, directly and indirectly, from firefighting activities;
(4) information on all types of firefighting activities, including inspection practices;
(5) technical information related to building construction, fire properties of materials, and similar information;
(6) information on fire prevention and control laws, systems, methods, techniques, and administrative structures used in foreign nations;
(7) information on the causes, behavior, and best method of control of other types of fire, including, but not limited to, forest fires, brush fires, fire underground, oil blow-out fires, and waterborne fires; and
(8) such other information and data as is deemed useful and applicable.

Ultimately, the National Fire Data System, which will include data from the fire services, from surveys such as that described above, and from other public and private sources, should provide a greatly improved source of fire information. However, such data collections can only provide retrospective information on past events. They cannot predict the future as new materials, new applications, and new construction techniques are introduced. We would like to anticipate hazards and recognize unsafe products before they appear as significant statistics in the fire record. This can best be accomplished by the laboratory measurement of pertinent properties of new materials, products and systems and the development and analysis of fire scenarios involving these items.

Another area where statistical surveys are of limited value is the infrequent catastrophic fire such as an aircraft crash fire (Lucha et al. 1975) or an oil refinery fire. Here the sample is too small for valid statistical analysis. However, the scenario approach can provide a useful tool for the identification of hazard.

3.3.2 Quantification of Hazard

Statistical data of the type described above serve to identify general areas where fires occur with significant frequency and severity, but they seldom provide the detailed information necessary to guide remedial action. For this purpose it may be necessary to know the precise materials or products which played significant roles in the fire, the nature of the ignition source, the growth pattern of the fire, the behavior of humans who may have been involved, the mechanism of injury or loss, and many other details of the incident. Such information is seldom available from routine fire reports and an in-depth investigation of selected incidents may be required (Buchbinder and Buchbinder, 1975).

For example, statistical data indicated the large number of injuries and deaths from fabric fires, leading to the 1967 Amendment to the Flammable Fabrics Act. However, the statistical data provided no guidance for the development of test methods and standards which would be more effective than the existing standard CS 191-53. To provide the needed information, the National Bureau of Standards established the Flammable Fabrics Accident Case and Testing System (FFACTS) (Vickers, 1977). This system, since taken over by the Consumers Product Safety Commission, provided in-depth investigations of more than 3,500 fire incidents involving flammable fabrics. Samples of the materials involved were obtained wherever possible for identification, laboratory testing, and sometimes accident simulation. More than 100 factors relating to a given incident could be coded for automated retrieval from the data bank (not all these data would be available from a given incident). Through analysis of this data file it was possible to identify such factors as the type of product most frequently involved, the sources of ignition, the types of fabrics, the age groups most frequently injured, and the cause of injury (burns, smoke or gas). It was then possible to develop test methods and standards directed to specific causes of loss.

A similar, but rather limited in depth, study of fires in which plastic materials played a significant role is now under way at the National Bureau of Standards. The data base is still too small to provide significant conclusions.

While in-depth investigation of real fires is an invaluable tool in quantifying hazards, it is limited by the incompleteness of the available information, the usual absence of trained observers, and the lack of quantitative data. Laboratory simulation of the fire can frequently be used to provide the missing information. Here the actual circumstances of the fire, as described in the accident scenario, are recreated as accurately as possible and the behavior of the resulting fire is compared to the scenario. Trained observers and extensive instrumentation may be used to provide a more objective and quantitative description of the event. A key question to be answered is "could the real fire have occurred in the manner postulated in the investigative scenario?".

3.3.3 Development of Test Method

With the identification and quantification of the hazard to be addressed, the development of a suitable test method to measure hazard potential can usually proceed in a straightforward manner. Malhotra (Malhotra, 1975) has listed the following attributes of a well-designed fire test method.

"(a) Environmental conditions: The test should reproduce the heating regime source, thermal feedback, oxygen supply, movement and dispersal of combustion products as is likely to be experienced in practice.

(b) Range of applicability: The environmental conditions should be capable of variation to increase the applicability of the test.

(c) Material representation: The modeling of the material should be such as to exclude effects of size, the presence of joints and junctions.

(d) Flexibility: The test should be capable of reproducing different orientations in which the product can be used.

(e) Reproducibility, repeatability and discrimination, should be of an acceptable level depending upon the nature of the test. A variance of 5 percent is satisfactory in most cases, in some even 10–15 percent is adequate.

(f) Ease of operation: Small tests should be capable of single-handed operation in no more than two hours, even complex tests should not take more than one day.

(g) Meaningful expression of results: The results should be expressed objectively in units which make comparison easy. Descriptive phraseology should be avoided."

We question the strict interpretation of item (c) since size must inevitably have an effect in real fires and the test method should give an indication of this effect. And if joints and junctions are characteristic of the mode of application of the product, the behavior of these under fire conditions is a legitimate objective of the test. Thus, in discussing the development of the Tunnel Test (essentially the current ASTM E 84), Steiner states (Steiner, 1961):

> "In the development of this test, the size of the specimen was aimed at the minimum which would reproduce actual behavior of surfaces under fire exposure conditions, which requires that the test surface be given the opportunity to develop conditions contributing to flame spread, such as distortion and separation of joints and to delamination. The larger the area, within limits, the more realistic is the behavior created by the fire exposure."

The reproducibility requirements suggested in Malhotra's item (e) also appear to be optimistic and unnecessarily restrictive. Such precision can be achieved under carefully controlled laboratory conditions; however, experience with test methods which simulate more closely the conditions which might be expected in a real fire indicates that a much larger variance is to be expected (Lee and Huggett, 1975). Such imprecise results can still be useful in separating materials into broad performance categories useful in fire safety regulations. Indeed, the present level of understanding of fire phenomena and the ability to translate such understanding into standards is so limited that more precise measurements would be of marginal value from a fire safety standpoint. This lack of precision does pose serious problems, however, when questions relating to legal compliance with standards come into play.

Finally, the fire performance of a specific material, product, or system may change during service life. Such factors as weathering, thermal stress, abrasion, etc.

can seriously alter the fire performance characteristics measured by tests which do not consider these factors. For example, water soluble additives or volatile components may be extracted or exuded and affect the material's response.

3.3.4 Development of Performance Standard

A final step in the technological portion of the process of reducing fire loss is the development of a standard or recommended practice which, when properly implemented, will reduce the probability of occurrence of the particular type of loss addressed. This is perhaps the most difficult step in the process. A good standard will consist of three parts: a test method by which the hazard potential of the material, product or system is to be measured; criteria which establish appropriate levels of performance for the given application (these may vary with the application for a given product); and sampling plans, inspection procedures and other auxiliary requirements by means of which the producer, consumer, and regulator can be assured that the product does indeed meet the requirements of the standards.

The test method selected must be appropriate to the hazard to be controlled. Thus a fire retarded cotton batting performed well in an open flame ignition test but underwent smoldering combustion when ignited by a cigarette. Hazard analysis had shown that smoldering combustion caused by a low energy ignition source was the most frequent hazard mode in mattress fires. On the other hand, blankets are not readily ignited by a cigarette, and hazard analysis demonstrated that a flaming ignition test method would be appropriate for the reduction of the frequency of blanket fires.

The establishment of suitable levels of performance is a critical step in the development of a standard. Too low a level may allow the continued existence of unacceptable hazards while too high a level will place an unreasonable burden on society in terms of increased cost and limitations on the choice of goods. Different levels of performance may be required for different applications. A lower level may be acceptable for single family dwellings where the occupants have easy access to exits and have a measure of control over their own activities, while a higher level of safety may be appropriate for a high-rise apartment building where egress is limited and the occupants may be exposed to hazards not of their own making. Similarly, a higher level of protection may be needed in institutional buildings where the occupants may be handicapped or physically confined than for buildings occupied by the normal population.

Formal cost-benefit analysis has not been applied to the setting of performance levels because of the complexity of the problem and the difficulty of obtaining adequate data (Dardis, 1975). Standards have usually been established on an intuitive basis, guided by accident experience, simulation experiments, and an estimate of the probable economic impact. For example, fire investigations and laboratory simulation experiments indicate that the "Pill Test" (DOC-FF-1-70) affords adequate performance assurance for floor coverings in the great majority of applications (Tu and Davis, 1976). A carpet which passes the test will not be easily ignited or spread the fire from a small ignition source. If a fire develops in a room, the floor covering will be one of the last fuel elements to become involved and will make a minor contribution to hazard. Economic studies indicated that the standard would have only a small impact on the floor covering industry. On the other hand, accident investigations identified one class of fires that would not be adequately controlled by the pill test standard; fires in which corridors and stairwells provided a path for fire spread through a building from a well established room fire. Full scale simulation experiments helped to quantify the hazard and guided the development of a suitable test method (Huggett, 1973). Tests of the floor coverings involved in several disastrous fires and in the simulation experiments suggested levels of performance adequate to control the hazard in various applications. Tests on a representative selection of floor covering from the market place gave an indication of the economic impact of the proposed levels of performance. The product of this extensive series of related studies was a recommended practice which, while possessing obvious limitations, is perhaps the most firmly grounded and thoroughly documented of any in the fire safety field (Benjamin and Adams, 1976).

Having established suitable levels of performance and a means of measuring performance, it is necessary for the supplier to be able to determine that his product meets the requirements of the standard. The consumer requires assurance that the product he purchases will perform as advertised, and the regulator must have a procedure for policing the market place. Since fire hazard tests are almost invariably destructive tests, the concept of 100% inspection is obviously inapplicable. Some form of quality control or statistical sampling plan is required. This problem has been addressed recently by Broussalian et al., and the following paragraphs are taken from the summary of their report (National Bureau of Standards, NBSIR 75-697, 1975).

"1. After identifying a hazard which he deems is in the public interest to reduce, the regulator begins by devising a technique for measuring some appropriate physical feature of the product which is suspected of causing injury. For example, in the case of fabrics, he may determine that the length of a char induced by a suitable ignition source correlates well with flammability and the risk of burn injury. Having made such a determination, the regulator is then in a position to specify an objective means of measurement and the setting of a proper standard. The latter consists of specifying a range of acceptable values of the product feature, e.g., flammability when it is measured as prescribed by the standard. It must be appreciated that any product, even though it meets the standard, still contains some residual level of risk.

2. Having established a "reasonable" standard by taking into account costs, benefits, and the residual risks which people are willing to assume, the regulator must further find an effective means of gaining compliance. The five which follow constitute his major available options. In the order of taking earlier and earlier action in the production process for the purpose of avoiding injury from unwanted

products they are the following: (I) processing of complaints about injurious and defective products leading; for example, to recall, to other administrative action, or to support of liability litigation; (II) off-shelf market place sampling of products leading to further action such as an investigation of the manufacturer's quality control procedures or to direct administrative legal action; (III) voluntary sampling or other quality control assurance provided by the producer; (IV) in-plant mandatory sampling prescribed by the regulator; and (V) prototype testing prior to production. Combinations of these can also be used."

A good fire safety standard will not, of itself, reduce fire loss. It must be implemented. This may be through voluntary compliance by producers, designers, and builders or, more frequently, through adoption as a mandatory standard, regulation, or code requirement by a government agency. Finally, there must be enforcement through inspection and possible legal procedures to assure compliance. It is apparent that with the development of suitable test methods and voluntary standards or recommended practices, the problem of the fire safety aspects of polymeric materials passes out of the hands of the technologists and into the hands of a much broader segment of society.

3.4 Classification of Fire Test Methods

Strictly speaking, a fire test method is a procedure which can be used to predict the performance of a material, product, structure, or system under a fire exposure condition that can reasonably be anticipated in the intended application. In practice, a great variety of physical observations and measurements have been referred to as fire test methods. They differ widely in purpose, scale, degree of sophistication, and other attributes, making systematic classification difficult.

Hilado (Hilado, 1973) divides test methods into two groups: research tests and acceptance tests. He points out that different methods may be applicable to materials in different physical states: gases, liquids, and solids in different states of aggregation. Further, test methods may measure different fire hazard characteristics such as:

ease of ignition
surface flame spread
heat release
smoke evolution
toxic gas formation
fire endurance

Acceptance tests may be further classified according to the intended end-use of the product which frequently requires compliance with specific regulations. Finally, tests are frequently classified according to size and designated by such descriptive and non-quantitative terms as large scale, full scale, subscale, small scale and laboratory scale. From a fundamental standpoint, this is the least useful type of classification but from a practical standpoint the size of the test may be very important.

Robertson (Robertson, 1975) classifies test methods as property tests and system tests, and makes a further distinction between nondestructive and destructive tests and between active and passive tests. Malhotra (loc. cit.) expands this list to include basic property tests, quality assurance tests, hazard assessment tests and ad hoc tests designed to deal with specific situations.

A multidimensional matrix would obviously be necessary to provide a detailed classification of the hundreds of test methods which have been described in published reports. In the following sections we discuss test methods under several of the headings suggested above without attempting a rigorous classification. Some methods may fit under more than one heading, thus a given test method may be considered from more than one point of view.

3.4.1 Research Tests

More properly, these should be referred to as research experiments rather than tests. Their purpose is to provide a better understanding of some particular aspect of fire behavior under well-defined conditions, rather than to predict product performance in a real fire.

Research tests are characterized by careful control of the environment of the experiment, extensive data collection (A full-scale bedroom fire experiment involved the recording of 192 channels of data at frequent intervals over a 15 minute period (Alpert et al., 1975)) and detailed analysis of the data aimed at developing a theoretical description of the phenomenon under study. The size of such experiments can vary widely, ranging from the micro-scale probing of a laboratory flame to one of the largest fire experiments ever undertaken, Operation Euroka, a 50 acre wildland fuel fire in Australia (Adams et al., 1973).

Scaling and modeling experiments constitute a special class of research experiments. Modeling experiments are usually smaller in size than the prototype. They are used because they may be less expensive, less hazardous, more reproducible and more amenable to precise measurement and analysis. The purpose of a modeling experiment is to develop rules for predicting the outcome of a prototype test from the results of a subscale test or experiment.

Successful modeling involves a great deal more than the mere reduction of the scale of the experiment. This is because the various output parameters of the fire (fuel consumption rate, convective energy transport, radiant flux, etc.) are not all related to the controllable input parameters (linear scale, fuel loading, ventilation, oxygen concentration, etc.) in the same functional way. Indeed, the complete scale modeling of a fire is impossible and all useful models represent cases of partial modeling where only selected aspects of the prototype fire which are of primary interest are modeled exactly. As Spalding (Spalding, 1963) has aptly said, "The central problem of partial modeling is to discern which modeling rules need not be obeyed, and to estimate the resulting errors in the predictions which are made."

A simple example of the difficulties encountered can be found in the observation that small flames are usually laminar in character while real fires are almost always turbulent. The transition from laminar to turbulent burning occurs when the characteristic scale of the fire exceeds a few inches. Obviously, a small laboratory flame can not be expected to model the combustion behavior of the same fuel in a turbulent fire environment.

Difficulties are also encountered in maintaining the ratio of radiative to convective heat transfer constant as the scale of the fire is changed. For example, the reradiation from a burning surface to the room depends on the temperature of the burning surface, which is relatively constant and not scale-dependent, while the energy flux from the flame to the surface increases markedly with scale.

In a more elaborate investigation of modeling techniques, it was shown theoretically that many of the characteristics of a large-scale fire are accurately modeled by a laboratory fire burning at an elevated pressure if the product of the cube of the length times the square of the pressure ($l^3\ p^2$) is the same for the model as for the prototype (de Ris et al., 1973). Both transient and steady state aspects of fires were modeled in a 2.15 m^3 pressure chamber at pressures up to 30 atm, with up to a ten-fold reduction in the characteristics scale length of the experiment over the atmospheric pressure prototype (Alpert, 1975). Not all aspects of the fire are modeled successfully by the pressure modeling technique, but it affords an excellent example of the application of partial modeling to the design of reduced scale experiments.

3.4.2 Property Tests

Property tests measure a property or performance characteristic of a material independent of the product or application in which it is to be used. They may be further subdivided into measurements of intrinsic properties, independent of sample geometry and test environment (e.g., heat of combustion, thermal conductivity, heat capacity) and measurements of performance properties which reflect an interaction with the test environment under carefully standardized conditions (e.g., autoignition temperature, oxygen index, specific optical density).

Test methods of this type are useful in product development and quality control, as well as in the establishment of specifications and in regulatory activities. They are less useful in setting fire hazard standards since the relationship between the measured property value and performance in an application under real fire exposure is seldom established. For example, the analytical determination of the concentration of fire retardant in a polymer composition may be a suitable method of quality control of the formulation process, but it tells little about how a product made from the material would perform in a fire.

Property tests are usually small in size, simple and inexpensive to carry out. In contrast to the research tests discussed in the previous section, they should be performed according to detailed and carefully documented procedures and the results should be capable of objective interpretation, free of any opportunity for subjective judgements. Since the results of such tests may be at issue in legal proceedings, such as disputes over specification acceptance of materials or compliance with a mandatory standard, the manner in which the test is conducted may take precedence over the significance of the results as a measure of hazard.

3.4.3 System Tests

In contrast to property tests, system tests emphasize interactions – interactions between material properties and the configuration in which the material is used, interactions between the various components which make up a system, and interactions between the system and its environment. They are designed to simulate the significant features of an anticipated fire exposure under application conditions. Thus, they may be used to predict performance and to provide the basis for effective hazard control standards.

The nature of the interactions which must be accounted for in systems tests is not always apparent at first glance. Thus, the fire performance of a carpet is found to be more dependent on the insulating properties of the pad and subfloor on which it is installed than on the combustion properties of these materials (Danyes and Quintiere, 1974). Obviously, a suitable system test for carpeting must measure the behavior of the entire floor assembly and not that of the individual components. Similarly, in testing the fire properties of interior furnishings in compartments, it is found that the rate of fire growth is affected significantly by the thermal inertia of the compartment walls even though they are of noncombustible material.

The world of real fires embraces an almost unlimited range of possible conditions. The following are some of the variables which can have significant effects on the outcome of a system test or a real fire:

- Ignition Source
  - Location
  - Temperature
  - Energy output
  - Duration
- Combustibles
  - Fuel Load
  - Composition
  - Availability
  - Configuration
  - Moisture content
- Ventilation
  - Natural
  - Forced

Wind velocity and direction
Venting during the fire
Confinement
Dimensions of compartment
Thermophysical properties of walls
Structural integrity

A single test represents only one point in this multi-dimensional space. It is seldom practical to conduct a sufficient number of tests to provide a reasonable sample of the fire conditions that can be anticipated. Therefore, it is extremely important to be able to generalize from the results of a particular test and predict the effects of limited variations in fire conditions. This is perhaps best accomplished through the scaling and modeling techniques referred to above, but it may also be accomplished empirically by testing at more than one level of the appropriate test variable. For example, the Flooring Radiant Panel Test (Benjamin and Adams, 1976) exposes the test specimen to a controlled range of radiation intensities. The fire performance of the material in a variety of fire exposures can be inferred from the results.

System tests are most useful as test methods to be referenced in fire hazard standards and codes. When properly designed, they combine a practical level of operability with a demonstrable relationship to performance in real fires.

3.4.4 Prototype Tests

Prototype tests expose the fully developed product, system, or structure under test to a fire environment that may reasonably be anticipated under conditions of actual use. By definition, they are full-scale tests although the actual size of the test will vary widely with the size of the item. For example, a prototype test of a child's sleeping garment might be carried out in a laboratory hood while a test of a pressurization system to control smoke movement might involve a complete multi-story building.

Prototype tests may be considered to be a limiting case of the system test where the conditions which determine performance are not merely simulated but are followed in exact detail. Prototype tests are apt to be very expensive, thus restricting the number of tests that can be conducted.

Circumstances sometimes provide opportunities for prototype tests (ad hoc tests) which would otherwise be prohibitively expensive. Thus, buildings scheduled for demolition have been utilized in a number of notable tests (See, for example, Shorter, 1961; Hill, Butler et al., 1973). Similarly, a bus which had suffered severe mechanical damage afforded an opportunity to conduct burnout tests on the vehicle's interior (Braun, 1975).

As in 3.4.3, a prototype test represents only a single point in the fire matrix and the ability to generalize from the results is severely limited. Tests are conducted on fully developed items, so the results are of limited value for development purposes. As a result of these limitations, prototype tests find only limited use in fire safety activities. Their chief value is in confirming the predictions drawn from property and system tests made during the development of the item under consideration. In situations where a very high level of fire safety is required, for example, in a manned satellite or an aircraft, prototype testing is the only means presently available to provide the required level of assurance.

3.4.5 Size and Scale of Tests

A great deal of recent discussion has centered on the appropriate size or scale of fire hazard tests (Benjamin, 1975; Wilson, 1976). The position has been advanced that only large-scale tests provide a reliable measure of hazard. In this connection, it is interesting to note that the tunnel test (ASTM E 84) has been referred to as a small-scale test although it requires a test specimen 25 ft. long. Much of the present confusion results from the lack of adequate and generally accepted definitions.

First, it is important to recognize the difference between *size* and *scale.* Size refers to the physical dimensions of the test assembly while scale refers to its size relative to that of the prototype. Prototype tests are, by definition, full-scale tests. They may be large or small, depending on the dimensions of the item under test. System tests may be full-scale tests or subscale tests depending on the scale on which the test item is simulated. Subscale tests are obviously smaller than their prototype and should make use of suitable modeling relationships to relate the results to the expected performance of the prototype. For operational convenience, the scale is frequently reduced to (and sometimes beyond) the limit where acceptable modeling is possible.

Small or laboratory size tests may be loosely defined as those which can be conducted in a conventional laboratory. Their characteristic dimensions will usually not exceed one meter and range downward to micro-scale property measurements. Large-size tests, on the other hand, may be defined as tests which require a specialized test structure or are conducted in the open.

Full-scale tests provide the most reliable measure of hazard potential, but their use is frequently impractical from the standpoint of cost and convenience. Subscale tests can provide a practical alternative if they are carefully designed to model correctly the essential features of the prototype test. Laboratory size property measurements are useful for development and control purposes, but great caution must be exercised in trying to relate them to hazard potential in real fire situations.

3.5 Fire Development

Most fire tests and standards treat fire as a quasi-steady phenomenon. Many test methods subject a sample of material of small fixed dimensions to an arbitrary energy pulse and record the response. Even in cases where the test method embodies a rate concept (ASTM E 162, NFPA 258, etc.), the results are usually used in building codes and standards in the form of integrated, non-time dependent material properties (flame spread index, specific optical density, heat of combustion, etc.). Thus, to specify that a material for a particular application shall have a maximum specific optical density not greater than 300 has little meaning unless the quantity of material, the volume of the compartment, and the rate at which the material becomes involved in the fire are known.

Codes and standards do give limited recognition to the time dependence of fire phenomena through the use of time ratings for structural components when subjected to a programmed temperature-time history (ASTM E 119, etc.). The Federal Aviation Administration has recently called attention to the concept of time dependent phenomena in fire standards through proposals to relate the allowable rates of smoke and toxic gas production to a "time to escape" (FAA, 1974; FAA, 1975).

Real fires are transient phenomena that follow a well-defined pattern. The course of a typical fire, unperturbed by outside intervention, is represented by the sketch in Figure 3.1. Here "t" is time and "I" is a measure of fire intensity, e.g., rate of fuel consumption, rate of heat release, rate of generation of combustion products, etc.

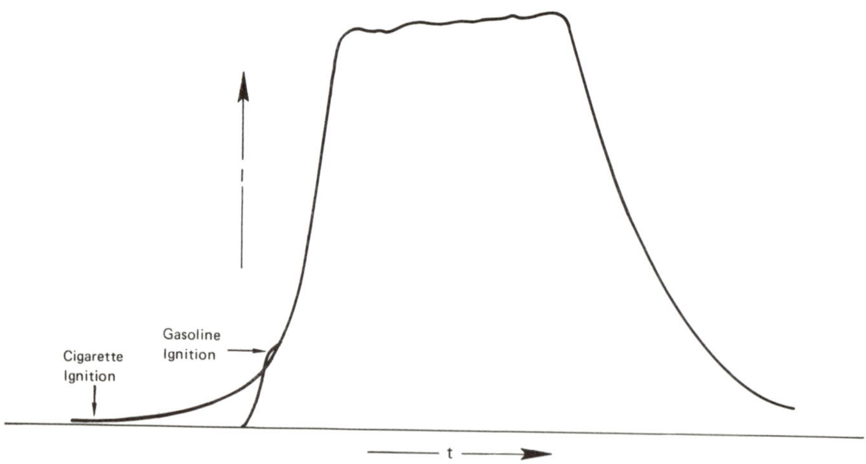

Figure 3.1. Typical fire development pattern.

Various stages in the fire development process are of significance in the control of fires and the limitation of loss. These stages are measured by the occurrence of critical events on the time line. Actually, two time lines must be considered. The first is defined by the progress of the fire and includes the time of ignition, the times to reach critical levels of temperature, smoke concentration, and concentration of toxic gases, the time to flashover, the time of structural failure, the time to the start of decay, and the time of extinction. The second is determined by the human (and mechanical) response to the fire and includes the time of detection, the time of activation of a fire extinguishing system, the time of evacuation, and the time of arrival of the fire department. In general, the extent of loss will depend on the relative sequence of events on these two time lines. The development of test methods and standards which give greater recognition to the time-dependent nature of the development of hazard should be a goal of future fire safety activities.

All fires start with an ignition event; a source of energy comes in contact with an easily ignited fuel (kindling fuel) in the presence of an adequate supply of oxygen and initiates an exothermic chemical reaction. If the rate of heat evolution is greater than the rate of heat dissipation to the surroundings, the temperature will increase and the fire will grow. Energy feedback to the fuel from the hot flame accelerates the growth exponentially.

The time of ignition is not uniquely defined with respect to the fire growth time line. In real fires, the exact time of ignition is seldom known. A low energy ignition source such as a lighted cigarette dropped on an upholstered chair would represent a point far to the left in Figure 3.1. The smoldering fire might grow slowly over a period of hours before breaking out into open flaming, followed by fire spread to other fuel elements and leading ultimately to flashover. In another case, a can of gasoline poured on the chair and ignited would lead to very rapid initial fire growth. However, once the gasoline has been consumed, further development of the fire will follow much the same course as in the previous example (Figure 3.1).

The rate of the later stages of fire growth is thus seen to be largely independent of the characteristics of the ignition source. The "time to flashover" is not an unambiguous measure of hazard potential.

The use of ignition source/kindling fuel combinations of increasing strength has been suggested recently for use in room burn experiments (ASTM, 1977).

It is apparent that increasing the energy output of the ignition source, provided the source is capable of initiating a sustained ignition, is equivalent to moving to the right along the axis of Figure 3.1. The test then becomes a measure of the strength of the ignition source needed to produce flashover in a given time as well as of the fire growth characteristics of the item under test.

The growth rate will depend on the geometry and combustion characteristics of the fuel elements, the dimensions of the fire compartment, ventilation, and other parameters relating to a specific fire. The initial growth of a fire is due to the spread of the fire over the surface of the fuel element first ignited, followed by spread to other fuel elements. The fuel consumption rate will be equal to the product of the fuel density, the burning area, and the linear regression rate of the burning surface.

The burning area will increase with time at a rate determined by the geometry of the fuel surface and the flame spread rate. Both the flame spread rate and the burning rate are functions of the fire intensity (energy feedback to the surface), accounting for the exponential growth of the fire. As the fire growth rate accelerates rapidly, a point will be reached where all of the readily available fuel surfaces are involved and further growth is limited. The fire then proceeds at a relatively constant rate determined either by the availability of fuel (fuel limited fire) or the availability of oxygen (ventilation limited fire) until fuel elements begin to be completely consumed. This process results in a decrease in fire intensity, and eventual extinction.

In a room or compartment, the stage at which all exposed fuel surfaces become involved is termed "flashover" (Waterman, 1968). At this point, conditions inside the compartment are clearly untenable and ad hoc efforts at fire fighting are unlikely to be effective. Flashover thus represents a significant stage in the fire development with respect to life safety. Not only will any occupants of the room of origin become casualties, but large volumes of flame and hot gases issuing from the flashed-over room will cause rapid involvement of other parts of the structure. Polymeric materials can play a very significant role in the development of flashover conditions since the fire, just prior to flashover, involves the principally polymeric interior furnishings and finishings of the room rather than the structural elements of the building.

While the period of fire growth leading to flashover is of primary importance with respect to the safety of persons, the post-flashover fire, spreading to other compartments and structures and ultimately leading to structural failure, makes a large contribution to property loss. This is reflected in the requirements found in many building codes for fire resistance measured in hours for structural members. The intensity and duration of the post-flashover fire is determined largely by the fuel loading and ventilation. An extensive discussion of the post-flashover fire is contained in a recent report by Babrauskas and Williamson (Babrauskas and Williamson, 1975).

Much effort has gone into the study of fully developed (post-flashover) fires and the development of test methods and standards for their control. In contrast, the period between ignition and flashover has received much less attention. With increased recognition of the role of interior furnishings and finishings in the early stages of fire growth, this area should receive much greater emphasis in the future. The role of polymeric materials is particularly important since polymeric products make up a significant and increasing part of the fire load in building and vehicle interiors. Improved test methods to characterize their contribution to fire growth and effective standards to guide and control their application are urgently needed.

CHAPTER 4

TEST METHODS

4.1 Introduction

The field of fire hazard testing of materials is being questioned and new concepts of performance testing are being discussed. This situation is due largely to two factors: an influx of new materials and products, principally based on synthetic polymers, into modern buildings and transportation vehicles and an increased emphasis on life safety in fire regulations. As a result, there has been a proliferation of new small scale test methods designed to evaluate materials and products. Most of these tests originated as ad hoc methods aimed at a convenient comparison of experimental materials. Their relevance to the performance of products under use conditions was seldom considered during their development. Nevertheless, many of these tests have achieved recognition as standard test methods by virtue of their endorsement by ASTM, NFPA, Underwriters Laboratories and other organizations.

The tests described and listed in this chapter are significant and/or widely used in the evaluation of fire performance of polymeric materials. A detailed description of each test is not feasible; however, each test is referenced (by ASTM numbers where possible) so that a complete description can be obtained if desired. In Chapter 5, these tests are discussed; in particular, the usefulness of current tests to evaluate the flammability characteristics of synthetic polymeric materials is addressed.

4.2 Tests for Ease of Ignition

Ease of ignition may be defined as the facility with which a material or its pyrolysis products can be ignited under given conditions of temperature, humidity, flow velocity and oxygen concentration. This characteristic provides a measure of fire hazard in that ignition of a combustible material is the first step in all fires and a material which ignites more easily than another will more readily contribute to the propagation and growth of fire. Some "measures" of ease of ignition are auto-ignition temperature, flash ignition temperature, and ignition time.

Almost any polymeric material can be made to ignite, given enough heat, oxygen, and time. Ease of ignition can, therefore, be measured by combining these elements under fixed conditions. The most simple tests show whether a specimen exposed to a combination of these elements under fixed conditions will or will not ignite.

Many tests provide a measure of both ease of ignition and surface flammability. Some tests measure surface flammability for materials that are easily ignited and measure the ease of ignition for materials that are difficult to ignite. For convenience in discussion, all tests that provide some measure of surface flammability are described in Section 4.3. Tests for ease of ignition are shown in Table 4-1.

Table 4-1. Tests for Ease of Ignition

TEST METHOD	SAMPLE SIZE	APPLICATION	TEST RESULTS
ASTM D-1929 (1)	3±0.5g	all organic materials	measures flash-ignition temperature and self-ignition temperature in a hot-convective air furnace
DOC FF-1-70 (2) (ASTM D-2859)	230x230 mm	textile floor coverings i.e. carpets	measures the char length from a mathenamine tablet ignition flame; specimen rated resistant to flammability if char does not extend beyond 178 mm from ignition point; passage of test reduces probability of carpet ignition
DOC FF-4-72 (3)	production unit	mattresses	measures char lengths from 18 cigarette ignition sources placed on mattress test surface; mattress passes test if char length on surface is not more than 5.1 cm in any direction from nearest cigarette
Proposed standard for Upholstered Furniture (PFF 6-76) (4)	varies depending on surface location to be tested; mockup assembly or finished upholstered furniture item	upholstered furniture	measures cigarette ignition resistance; individual cigarette test locations pass the test if the char length is not more than 7.5 cm in any direction
ASTM E-136	38 mm x 51 mm x 38 mm	all organic and inorganic materials	materials reported as noncombustible if meet temperature rise, flaming and weight loss criteria
ASTM D-229	127 x 13 mm x thickness of specimen (>1 mm)	electrical insulation materials normally manufactured in flat sheet or plate form	measures both burning rate and ignitibility, reports ignition and burning times

(Continued)

Table 4-1. Tests for Ease of Ignition (Continued)

TEST METHOD	SAMPLE SIZE	APPLICATION	TEST RESULTS
Federal Test Method Std. No. 406 (5)			
Method 2023	5 in x 0.5 in x 0.5 in	all plastics	measures the time to ignite the specimen (sec), maximum distance of flame travel, and burning time of specimen centered in a heating coil
Method 4011	rectangular specimen 0.125 + 0.010 in thick	electrical insulating materials	measures ability to resist the action of an arc of high voltage and low current close to the insulation surface to form a conducting path therein; reports arc-resistance time
UL Ignition Tests (6)		plastics used in electrical appliances	
Ignition Temperature Test	rectangular solids 1/4 x 1/8 in. to dust size particles		measures minimum furnace temperature to cause flaming or glowing
High Voltage Arc Resistance (Tracking) Test	6" x 1/2" x 1/16" or 1/8" or 1/4" thick		measures ignitibility to arc across specimen; (measures ignitibility while electrodes separated and arc extinguished)
Hot Wire Ignition Test	3"x1/2"x1/6" or 1/8" or 1/4" thick		sample wrapped in electrically heated wire; observations include ignition, dripping (flaming or non-flaming) free or restricted burning

(Concluded)

4.3 Flame Spread Tests

Flame spread may be defined as the rate of travel of the flame front. This characteristic is a function of both the chemical composition of the material and the physical conditions of the test (sample orientation and dimensions, environmental conditions such as air velocity, oxygen concentration, external radiant heat flux, etc.). Flame spread rate provides a measurement of fire hazard in that it controls the time after ignition when the fire reaches a dangerous size. Flame spread tests have been the most common measure of flammability in the past and consequently many flame spread tests exist. Table 4-2 lists many of these tests.

4.4 Rate of Heat Release Tests

The conventional calorimeter, (e.g., the Parr Bomb) measures the thermochemical quantity of heat released when a known small quantity of fuel is completely burned in oxygen and the products of combustion are cooled to the initial fuel-air mixture temperature. The heat-release rate calorimeter (HRRC) introduces a time-scale over which the heat is released, and aims to define the rate at which a material exposed to fire conditions contributes heat to the fire. This characteristic provides a measure of fire hazard in that a material which burns with the evolution of little heat per unit time will contribute appreciably less to fire growth than a material that generates large amounts of heat per unit time. Table 4-3 lists some rate of heat release tests.

4.5 Tests for Oxygen Requirements

A variety of research tests have been used in which the burning of a test specimen is observed in atmospheres containing varying amounts of oxygen. One of these tests (limiting oxygen index) has achieved the status of a standard test method and is listed in Table 4-4.

4.6 Tests for Ease of Extinguishment

While test procedures exist to evaluate effectiveness of portable extinguishers on certain standardized fires (e.g., wood pellets, n-heptane pools), the inverse procedure of using a standardized extinguisher to test a variety of materials has never been developed. Thus, there are no standardized test methods to determine the best extinguisher for a particular polymeric combustible, or conversely to determine the relative ease of extinguishment of a series of polymeric combustibles by a given extinguisher.

Full-scale tests (or partial full-scale tests) to determine the ability of automatic sprinklers or foam applicators to control fires of specific materials are frequently performed at fire tests centers such as Factory Mutual or Underwriters Laboratories, but these tests have not developed into any standardized tests.

Table 4-2. Flame Spread Tests

TEST METHOD	SAMPLE SIZE	APPLICATION	TEST RESULTS
ASTM E-84 Tunnel Test	20 in x 25 ft x actual thickness	interior finish materials (e.g. paneling, ceiling tile, etc.) sheathing, thermal/acoustic insulation	primarily measures surface flame spread; values range from 0 for "non-combustibles" such as asbestos cement board to 100 for red oak & higher for other materials; also measures smoke generation and fuel contribution
ASTM E-162 Radiant Panel	6 in x 18 in x 1 in max.thickness	generally used as a screening test; may be used for applications of E-84	measures flame spread factor and temperature rise; values originally intended to be similar to ASTM E-84
ASTM E-286	8 ft x 13.75 in x actual thickness	building materials; generally used in research and development	similar to ASTM E-84; measures surface flame spread, smoke density and heat contribution
Flooring Radiant Panel Test[7]	100 cm long	floor covering systems installed in buildings corridors and exitways	measures minimum radiant flux necessary to sustain flame propagation on the flooring surface, acceptance criteria: ≥ 0.25 W/cm^2 for residential and commercial occupancies, ≥ 0.50 W/cm^2 for institutional
Federal Test Method Std. No. 501[8]			
Method 6411	31.5 in x 7 in	resilient nontextile floor coverings	measures combustion plus ignition time, char length, flame length and smoke density
Method 6421	18 in x 6 in	resilient nontextile floor coverings	measures flame spread index under radiant heating (similar to ASTM E-162)

(Continued)

Table 4-2. Flame Spread Tests (Continued)

TEST METHOD	SAMPLE SIZE	APPLICATION	TEST RESULTS
ASTM D-635	125 mm x 12.5 mm x supplied thickness	laboratory screening test of relative flammability of self-supporting plastics in form of bars	measures horizontal burning rate when flame proceeds 100 mm, or time and extent of burning otherwise
ASTM D-568	25 mm wide x 45 cm length	plastics in the form of flexible, thin sheets or films	measures vertical burning rate, or time and extent of burning otherwise
ASTM D-1433	76 mm wide x 228 mm length	flexible plastics in the form of film or thin sheeting	measures burning rate on specimen mounted at a 45 degree angle, or time and extent of burning otherwise, including dripping, flaming, etc.
ASTM D-1692	150 mm x 50 mm x up to 13 mm thick	laboratory screening procedure for rigid or flexible cellular plastics	measures horizontal rate of burning or extent of burning
Federal Test Method Std. No. 406[9]			
Method 2021	5 in x 0.5 in by normal thickness	rigid plastics in form of sheets or molded bars	measures horizontal burning rate when flame proceeds four inches; or length burned otherwise (similar to ASTM D-635)
Method 2022	18 in x 1 in x thickness of sheet	thin plastic sheets or films	measures time required for the flame to either extinguish itself or to completely burn specimen, also reports area of specimen burned (similar to ASTM D-568)

(Continued)

Table 4-2. Flame Spread Tests (Continued)

TEST METHOD	SAMPLE SIZE	APPLICATION	TEST RESULTS
ASTM E-108 (UL 790 for limits)	1 m x 1.3 m 1 m x 4.0 m	roof coverings	measures response of roof deck to intermittent flame, measures flame spread distance, and subjects the roof deck to burning brands. Roof covering materials are rated Class A (best), Class B and Class C depending on performance in these tests.
UL 94 (10)	127 mm x 12.7 mm 152 mm x 50.8 mm	plastic parts in devices and appliances	measures burning rate in the horizontal and vertical configurations depending on classification. Test essentially employs ASTM D-635 and D-1692 for horizontal burning test and ASTM E-162 Radiant Panel Test. Also includes an ignition test.
UL 44 (11)	depends on paragraph	rubber-insulated wires and cables	measures burning extent and time, ignition of cotton underneath from dripping
UL 83 (12) (flame test)	approx. 46 cm	thermoplastic-insulated wires	measures flame spread and flaming resistance of vertically mounted wire specimen
IEEE 383 (13) (flame test)	multiple cables mounted in a tray 3 in deep x 12 in wide x 8 ft long	Class IE electric cables	cables which propagate the flame and burn the total height of the tray above the flame source fail the test. Cables which self-extinguish when the flame source is removed or burn out pass the test
FM 4922 (14)	7.2 m	air and fume handling ducts	to pass the test flaming must not proceed from the ignition fire to the other end of the duct, interior duct temperatures are limited to 1000 ^{o}F during the test and maximum one-minute heat release rate of duct interior must not exceed 200 $Btu/ft^2/min$

(Continued)

Table 4-2. Flame Spread Tests (Continued)

TEST METHOD	SAMPLE SIZE	APPLICATION	TEST RESULTS
NFPA 701[15]	Small scale test 7 cm x 25 cm Large-scale test 12.7 cm x 2.1 m 0.6 m x 2.1 m	flame resistance textiles and films; used in the interior furnishings of buildings and transport facilities, protective clothing, and for protective outdoor coverings such as tarpaulins and tents	test limits specimen flaming time, vertical flame spread, and flaming of drippings
Federal Std.[16] No. 191		Textiles	
5900	7 x 10 inches	Flame resistance of cloth-horizontal	ignition source is a cup containing 0.3 ml alcohol. Test measures distance traveled by the flame
5903	2-3/4 x 12 in	Flame resistance of cloth-vertical	measures resistance of cloth to flame and glow propagation, and tendency to char
5904	any size which contains an edge	Vertical field test for flame resistance of cloth	measures flaming time and char length
5906	4-1/2 x 12-1/2 in	burning rate of cloth, horizontal	ignition by specified gas burner. Rate of burning reported in inches per minute or per second to nearest 0.5 inch.
5908 See also ASTM D-125	2 x 6 inches	flammability of cloth 45° angle	rough measure of rate of burning or flashing and ease of ignition. Test is the standard for all fabrics (wearing apparel) of commerce.
DOC-FF-3-71 amended [J.AATCC 4(9) (Sept. 1972)]	8.9 x 25.4 cm (3.5 x 10 in)	children's sleepwear size 0-6X	measures char length in test specimen and residual flame time. (sampling plan, and other requirements are included in the standard)

Table 4-2. Flame Spread Tests (Continued)

TEST METHOD	SAMPLE SIZE	APPLICATION	TEST RESULTS
DOC-FF-5-74	8.9 x 25.4 cm (3.5 x 10 in)	children's sleepwear size 7-14	same test as DOC-FF-3-71 (Criteria for passing differ)
Motor Vehicle (17) Safety Std. No. 302	4 x 14 in	flammability of interior materials used in passenger cars	material tested in a horizontal position. Ignited with a flame. Test measures the time when the flame from the burning specimen reaches a point 1-1/2 inches from the clamped end of the specimen. This test is essentially the same as SAEJ-369, GM Test Method 32-12, and Chrysler LP-463KC-13-01

(Concluded)

Table 4-3. Rate of Heat Release Tests

TEST METHOD	SAMPLE SIZE	APPLICATION	TEST RESULTS
ASTM E-84	20 in x 25 ft x actual thickness	interior finish materials (e.g. paneling, ceiling tile, etc) sheathing, thermal/ acoustic insulation	reports total heat release, not rate of, and provides information relative to red oak only
ASTM E-162 Radiant Panel	6 in x 18 in x 1 in max. thickness	same as ASTM E-84	same as ASTM E-84
NBS Calorimeter (18)	11.4 cm x 15.2 cm	same as ASTM E-84	measures heat release rate as a function of time; typically reports peak, 1-min, 5-min and 10-min averages, for heat release rates
OSU Calorimeter (19)	25.4 cm x 25.4 cm	same as ASTM E-84	same as NBS Calorimeter
FM Calorimeter (20)	4.5 ft x 5 ft x actual thickness	composite roofing assemblies	measures heat release rate as a function of time; roof classification depends on peak rate
SRI Calorimeter (21)	0.46 m x 0.61 m	same as ASTM E-84, plus model wall and other construction assemblies	same as NBS Calorimeter
FPL Calorimeter (22)	0.46 m x 0.46 m	same as ASTM E-84, plus models of construction assemblies	same as NBS Calorimeter

Table 4-4. Tests for Oxygen Requirement

TEST METHOD	SAMPLE SIZE	APPLICATION	TEST RESULTS
ASTM D-2863 (LOI)	Depends on plastic form e.g. self-supporting 6.5 mm x 3.0 mm x 70-150 mm	Various forms of plastics including film and cellular plastic	measures the oxygen index (minimum concentration of oxygen in flowing O_2/N_2 mixture to just support flaming combustion; also descriptions of charring, dripping, bending, etc.

Table 4-5. Tests for Smoke Evolution

TEST METHOD	SAMPLE SIZE	APPLICATION	TEST RESULTS
NFPA 258 [23]	3 in x 3 in x 1 in max. thickness	all organic and inorganic solid materials and assemblies	measures maximum optical smoke density under flaming and smoldering conditions under an external radiant flux of 2.5 W/cm^2
ASTM D-2843	1 in x 1 in x 1/4 in thick	all plastics	measures maximum smoke density in percent light absorption, smoke density rating, for flaming combustion only
ASTM E-84	20 in x 25 ft x actual thickness	interior finish materials (e.g. paneling, ceiling tile, etc.) sheathing, thermal/acoustic insulation	measures light absorption by the smoke relative to red oak

4.7 Tests for Smoke Evolution

Smoke density may be defined as the degree of light or sight obscuration produced by the smoke from a burning material under given conditions of combustion. This characteristic provides a measure of fire hazard, in that an occupant has a better chance of escaping from a burning structure if he can see the exit, and a firefighter has a better chance of putting out the fire if visibility is adequate. Some measures of smoke density are: degree of light absorption, specific optical density, and smoke development factor.

Tests for smoke evolution generally involve measurement of the fraction of light absorbed or obstructed by smoke evolved from a decomposing or burning material. The degree of obscuration is a function of the number and size of particles, refractive index, light scattering, rate of movement, extent of ventilation, and distance through which light must travel. Table 4-5 lists tests for smoke evolution.

4.8 Tests for Toxic Gas Emission

Tests for the toxicity of a material's pyrolysis and combustion products generally fall into two types: those concerned with identification and analysis of the chemical compounds in the gaseous combustion products, and those concerned with studying the effects of these gases on laboratory animals. There are many laboratories in this country and throughout the world using various test apparatuses and procedures for toxicological studies; however, to date, no test method has emerged as the accepted standard for toxic gas emission determination. For a more detailed discussion of toxicity tests, see Volume 3 of this series.

4.9 Tests for Fire Endurance

Fire endurance may be defined as the resistance offered by the material to the passage of fire, normal to the exposed surface over which the flame spread is measured. This characteristic provides a measure of fire hazard in that, other things being equal, a material which will contain a fire represents more protection than one which will give way. Some measures of fire endurance are penetration time and resistance rating.

Most fire endurance tests are primarily concerned with the complete system rather than with individual materials, because it is widely recognized that the performance of individual materials that comprise the system are not necessarily indicative of the performance of the system as a whole.

Tests for fire endurance have been a vital part of fire safety testing for many years. Table 4-6 lists the major fire endurance tests.

4.10 Full-Scale Tests

Full scale fire tests are designed to reproduce actual fire scenarios under controlled and measured conditions. Generally, they can be divided into three types: (1) corner, (2) compartment, and (3) corridor tests. Compartment-and-corridor

Table 4-6. Tests for Fire Endurance

TEST METHOD	SAMPLE SIZE	APPLICATION	TEST RESULTS
ASTM E-119	Walls $\geq$ 100 ft^2 Floor/ceiling $\geq$ 180 ft^2	Complete walls Floors/ceilings	Determines ratings in hours. Failure criteria: flame penetration on unexposed face, back surface average (single point) temperature rise $\geq$ 250 °F (325 °F), structural collapse
	Column length $\geq$ 8 ft Beam/girder length $\geq$ 12 ft	Individual structural elements (e.g.columns, beams, girders, etc)	As above plus steel temperature measurement
ASTM E-152	Full-scale door and door frame installations, test assembly $\geq$ 100 ft^2	doors	determines ratings in hours: no flame penetration on unexposed face. Door must also remain in place during hose-stream test immediately following fire endurance test
ASTM E-163	Windows and shutters installed; test assembly $\geq$ 100 ft^2	windows	must withstand fire endurance test for 45 min. and hose-stream test that follows

Table 4-7. Full-Scale Fire Tests

TEST METHOD	SAMPLE SIZE	APPLICATION	TEST RESULTS
ASTM E-603	8'x8', 8'x12', 10'x10';with 8' ceiling	rooms(materials, products, systems)	time to flashover, total rate of smoke production, extent of flame spread for a low energy ignition source, size of ignition source required to produce flashover
Factory Mutual Building Corner Fire Test	Walls 50'x24'-9" 37'-9"x24'-9" Roof/ceiling 20-26 gauge corrugated steel	interior finish materials	material tested is considered acceptable if it does not produce a self-propagating fire within the limits of the structure as evidenced by flaming or material damage
UN.California-Berkeley-Corner Test	Corner 6'x6'x8' high Ceiling 6'x6'	interior finish materials	measures extent of flame travel and damage

tests, or room-and-corridor tests, could be considered as a fourth type, but these tend to be either compartment tests with additional information on the adjacent corridor, or corridor tests with an adjacent compartment used as the fire source. Table 4-7 lists commonly used full-scale tests.

CHAPTER 5

DISCUSSION OF TEST METHODS

5.1 Introduction

Laboratory tests are designed to evaluate some aspect of the fire performance of a material or assembly in a reproducible simulation of some real life situation. These "small-scale" tests are useful, since they are economical in comparison to full-scale test replications of rooms, corridors, buildings, etc. The major problem, however, is that there is a general lack of correlation between small-scale laboratory test results and real-life large-scale fire behavior. In a recent paper (Malhotra, 1974) the author states:

> "The early tests were devised with only a limited understanding of laws governing fire growth and severity and while in the last decade the knowledge has improved, the material technology has progressed at a tremendously fast rate with demands for more and more fire tests outstripping the gains in knowledge. It is, therefore, not surprising that the ratio of satisfactory to unsatisfactory tests has deteriorated."

Test methods used today were developed on the experience that test designers were able to relate to the specific problem (e.g., ASTM E 84, ASTM E 162, ASTM E 119). At the time many of these tests were developed, this experience was limited to fires involving mainly cellulosic materials. Primarily, these tests ranked materials in reference to a given cellulosic material (e.g., red oak) and through experiences of the test designers provided a measure of the hazard posed by these materials in end-use situations. However, through technological advances and extended use of synthetic materials, the tests that had been doing an adequate job regulating materials may no longer be valid.

In this chapter, current tests employed to measure the flammability characteristics of materials are discussed; in particular, their usefulness in evaluating synthetic polymeric materials is addressed.

5.2 Ignition Tests

Ignition of a combustible material is the first stage in any fire. Once the fire is started, ignition delay times of other materials, coupled with flame spread, affects the rate of fire spread and growth. Hence, control of ignition is crucial to fire prevention and control.

Most polymeric materials will ignite and burn if sufficiently heated in the presence of air or other atmosphere containing sufficient oxygen. Consequently, from the point of view of fire safety, it is desirable to know how long it takes a particular polymer to ignite, and whether or not it will ignite when exposed to a given oxygen containing environment and a given heating rate.

The ignition of a polymer is an extremely complex process depending on heat and mass transfer, thermal degradation, and chemical reactions. Since it is generally agreed that for most polymers, flaming ignition occurs in the gas phase above the polymer surface and involves an exothermic reaction between oxygen and the fuel's pyrolysis products, the degree of air motion or turbulence, as well as access to oxygen, are important. Details of heat transfer from the ignition source to the polymeric material will also influence the ignition process. This heat transfer may be some combination of conduction, convection and radiation.

The thickness and thermal properties of a material are vital in determining the time required to achieve ignition, given specific heat flux and environmental conditions. The time to igntion for a "thermally thick" specimen (see Sec. 3.3.3 of Volume 4) is independent of the thickness and controlled by the "thermal inertia," which is the product of the thermal conductivity and the heat capacity per unit volume. For a "thermally thin" specimen, the time to ignition is proportional to the product of the thickness and the heat capacity per unit volume (fabrics are generally in this category). Whether the specimen behaves in a "thermally thick" or "thermally thin" manner depends not only on the physical thickness, but also on the heating rate, the heating time and the "thermal diffusivity," which is the ratio of thermal conductivity to heat capacity per unit volume.

In the case of a thin flammable material (carpet, paneling, etc.) in thermal contact with an underlying material, the thermal properties of the underlying material can strongly influence the ignitability by the degree to which the underlying material acts as a heat sink.

Configuration and orientation of the polymeric material can also be of great importance. Ignition tends to occur more readily in a crevice or fold, at an edge or corner, etc., rather than in the middle of a flat surface.

The complexities of ignition phenomena make it impossible to devise a single test that will determine the ease of ignition of different materials for a variety of end-use situations. It has already been pointed out that ignitability will depend on specimen thickness and geometry, the type and method of application of the ignition source, ventilation and environmental conditions. Consequently, most ignition tests are of limited value. Many of the ignition tests currently employed (e.g., ASTM D 1929, ASTM D 229, UL ignition temperature test, UL hot wire ignition test, etc.), expose a specimen to heated air and measure the temperature (and/or time) required for ignition. This so-called "ignition temperature" is not an intrinsic property of the material, but depends strongly on ventilation rate, test specimen size and obviously on the heating conditions in the test. Yet these "ignition temperatures" are frequently reported and used to rank materials.

Another series of ignition tests developed at Underwriters Laboratories to evaluate the ease of ignition of materials intended for electrical applications involves subjecting the specimens to a high-voltage or high-current arc. Generally these tests are "go, no-go"tests, in which the specimens either do or do not ignite. It is not clear whether these tests represent a realistic service condition; therefore, results must be interpreted with caution.

Two recently developed ignition tests designed to simulate cigarette ignition have avoided the complexities mentioned above by employing an everyday ignition source, a cigarette with materials in a specific application, either as a mattress or as covering and padding for upholstered furniture. These tests recognize that this ignition source frequently initiates fire in these two items. Hence, tests were developed to measure the hazard posed by cigarette ignition of such items. Despite the simplicity of these tests, once the optimum arrangement and location for placing the ignition source was determined, the results are a good indication of the hazard for these common occurrences. Obviously, these tests do not simulate fire performance of these materials when exposed to a flaming ignition source.

There seems to be little current work going on developing additional ignition tests. Due to the complexities of the ignition phenomena, it would seem that future ignition tests should be directed towards determining whether or not ignition occurs from a given source, e.g., match flame, grease pan kitchen fire, etc., to a given item, e.g., drapes, kitchen cabinets, etc. This approach, determining ignitibility in a specific instance, could be based on fire statistics data of the most probable "ignition source – ignited item" combinations and would give an accurate measure of the ease of ignition under the specific circumstances of the test. An alternate approach would be to measure some ignition property, e.g., ignition time, in a carefully controlled experiment, recognizing that the quantity measured is not an intrinsic property of the material, and attempt to establish some relationship of this "property" to fire safety in actual situations.

5.3 Flame Spread Tests

Unless a person is in intimate contact with the ignited item, a fire is not apt to do much harm until it has grown by spreading some distance from the point of ignition. The rate of flame spread is very important in the history of a fire, because, as noted earlier, it controls the time after ignition when the fire reaches a dangerous size. The "dangerous size" relates to the rate of heat release, to the rate of generation of toxic and smoky products, or to the difficulty of extinguishment. The ability to detect, fight, or escape from the fire depends on the time before the fire reaches a dangerous size and hence, on the spread rate.

The propagation of a flame over a combustible solid is an extremely complex process. A large body of test methods have evolved over the past thirty years designed to measure the rate of flame spread. Many of these tests were developed without allowing for the numerous factors influencing the flame spread rate. Efforts have been largely fragmented and the test methods developed yield results that are generally not consistent and do not adequately reflect behavior in actual fires.

Results from many experimental investigations have been summarized

(Friedman, 1968; Magee and McAlvey, 1971). It is indicated that the flame spread velocity is affected by many parameters such as:

Physical and geometrical parameters including
- orientation of surface
- direction of propagation
- thickness of specimen
- specimen size
- surface roughness
- presence of sharp edges or crevices
- initial fuel temperature
- environmental pressures
- flow velocity of environment
- external radiant flux
- humidity

Chemical parameters including
- composition of solid
- composition of atmosphere

It is essential that individuals involved with test method development and persons responsible for selecting tests as a basis for material specifications recognize how strongly some of these physical and geometrical parameters affect the flame spread rate. For example, vertical flame spread is a continuously accelerating process; for small specimens the flame spread rate upward is at least an order of magnitude faster than vertical downward or horizontal flame spread (Magee and McAlevy, loc. cit.). Also, horizontal flame spread over specimens with exposed edges occurs approximately five times faster than when the edges are inhibited. Raising the initial temperature of polymethylmethacrylate from 25°C to 150°C doubles the horizontal flame spread velocity. Increasing the specimen thickness for "thermally thin" specimens lowers the flame spread rate proportionally; since for "thermally thin" specimens the flame spread velocity varies inversely with the specimen thickness (Magee and McAlevy, loc. cit.). These examples show how important an understanding of the factors influencing flame spread is when either developing or selecting an appropriate flame spread test.

Since, to propagate the flame, heat must travel ahead (possibly augmented by radiation from the surroundings) to the unignited material, certain heat transfer modes must be involved. However, the flame spread rate is affected by and changes with variations in the magnitude and relative importance of conduction or convection in the gas phase, conduction in the condensed phase, and radiation in the gas phase. Thus, tests which presume to establish relative flame spread characteristics of polymeric materials in a definitive manner are valid only for the particular conditions of the test and may give misleading results if extrapolated to other conditions.

Small-scale laboratory screening tests in this category include ASTM D 635, D 1692, D 3014, D 568 and D 1433. All these tests claim usefulness for determining the relative or comparative burning characteristics of materials. Actually they compare the behavior of materials only under the unique conditions of the test. Changing the specimen size, thickness, orientation or initial temperature, or changing any of the environmental conditions, changes the magnitudes and relative importance of the heat transfer modes involved and thus changes the flame spread velocities and most likely the relative flame spread characteristics of various materials. Also, since none of these tests simulates the energy feedback from the surroundings that occurs in real fires, the magnitudes of the spreading flame velocities obtained are unrealistically low. It cannot be assumed that any of these small-scale laboratory screening tests meaningfully "rank" the flame spread characteristics of materials in any other size, geometry, and "fire" condition than the precise one tested.

When a material is used as an interior wall or ceiling finish, it will affect the fire hazard at the place of use according to the extent to which it permits spread of flame over its surface, contributes fuel to the fire, or generates smoke and toxic gases when burning. In the United States, the Steiner Tunnel Test (ASTM E 84) is the most widely used procedure to measure the potential hazard of room lining materials. While this test reports measurements of fuel contribution, smoke density and rate of flame spread, this procedure is primarily used to determine flame spread. Despite considerable discussion and objection regarding technical details of the test, the ratings resulting from this method have widespread use. They are used in several national model building codes, many local codes, and by various regulatory bodies primarily as means for limiting the use of combustible interior finish materials in buildings.

This test, developed by Steiner in the early 1940's, was predicated on measuring the hazard resulting from fire propagating up a wall or along a ceiling in a room or corridor. The need for an interior wall finish material test was demonstrated by several fires in which the interior finish was the key factor in life loss. The test was developed with the intent of reproducing conditions consistent with field experience. Such conditions were created by adjustments of the fuel and air supply until the fire hazard properties of known materials were properly ranked. At that time, experience was solely with wood and similar cellulosic materials, and the test was developed to provide a basis for comparison of the flame spread hazard of these materials.

This test seemed adequate until the development and use of certain types of synthetic polymeric materials such as low density polyurethane and polystyrene foams, and several types of thermo-plastics. The test method does not provide a satisfactory rating for materials which soften, melt and drip or are of very low density.

Thermoplastic foams rapidly melt in the vicinity of the impinging gas flame and pull away from the roof support. This physical phenomenon prevents flame spread down the tunnel and consequently the material is given a low flame spread rating. Meanwhile, the drippings continue to burn on the tunnel floor, yet this obvious hazard in real life situations is not reflected in the calculation of flame spread classification (fsc), although it is recorded as an observation in the test report.

The low density of these foams also results in a misleading indication of fire hazard in this test. Due to low density and hence a low thermal conductivity, only a thin surface layer of foam is heated by the hot gases in the tunnel. Thus, the rate of production of combustible pyrolysis gases is low and the gases are swept away before burning. In a room fire however, these combustible pyrolysis gases would be confined and eventually burn, augmenting the fire intensity in the room. Thus some materials burn in a very hazardous manner in the real world whereas the test indicates that they should not.

Therefore, ASTM E 84 should not be employed as the sole criterion for fire hazard of all interior finish materials until such time as the test is modified to correct for the shortcomings mentioned above.

Work on the utility of this test for synthetic polymers, especially foams, is underway. For example, in a series of tests for the Society of the Plastic Industry (Christian and Waterman, 1970), a corridor was employed to evaluate the correlation of the ASTM E 84 flame spread classification for materials used on the walls and ceiling with fire behavior. The fire was started in a 10 X 15 ft (3.1 X4.6 m) room with an 8 ft (2.4 m) high ceiling adjacent to the corridor. The time for various linings to burn the length of the corridor was measured. It was concluded that "It is clear that placement of the materials in the order of ascending tunnel test flame spread ratings does not quite place them in the order of increasing flame spread rate or decreasing time in the full-scale corridor."

In a recent series of tests conducted at Underwriters Laboratories (Castino, 1975), the flashover characteristics of rooms lined with low density foamed plastics and other common building materials used for interior finishes were investigated. Various ignition sources were employed. It was concluded that:

> "Total incident heat flux levels in the test flame area of the 25 ft tunnel test are comparable to heat fluxes in the flame area of a 20 lb burning wood crib ignition source in a corner geometry."
>
> A major conclusion of this report is: "The flame spread classification of materials developed in the standard 25 ft tunnel test corresponds with the performance of those materials in corridor, corner and vertical-wall full-scale building geometry tests."

This conclusion was based on a criterion for acceptance; namely the determination of whether the room reached flashover conditions as evidence by a ceiling maximum temperature greater than 649°C (1200°F) or whether full ceiling involvement occurred.

However, if the time to flashover is employed as the criterion for hazard, then a different conclusion would be reached. Specifically, studying Table 5-1, constructed from data in the Castino-Underwriters Laboratories report, one readily sees little correlation of time to flashover with flame spread classification – in fact, for materials with the same classification, flashover times can vary by an order of magnitude. Also these data indicate that cellulosic materials used as room linings produce a different (and seemingly lesser) degree of hazard than polymeric foams having the same flame spread classification.

Table 5-1. Time to Full Ceiling Involvement (with 20 lb. crib)

Sample	Material	ASTM E 84 FSC	Time
	Room Tests		
S	Polyisocyanurate foam, FR	22	1:20
A	Polyisocyanurate foam, FR	23	1:40
G	Treated plywood	23	--*
H	Untreated plywood	178	4:20
	Corner Tests		
J	Untreated fiber board	54	10:00
Q	Polyurethane foam, FR	59	1:22
AD	Polyurethane foam, FR	75	0:40
AE	Red Oak	100	8:45
U	Untreated particle board	156	3:15
K	Untreated plywood	159	4:00
AC	Polyisocyanurate foam, FR	364	1:25
D	Polyurethane foam, FR	925	1:30
O	Polyurethane foam, FR	1735	0"22

* Not reached in 20 minutes
FR - flame retardant added.

The examples shown in Table 5-1 show that the ASTM E 84 test does not consistently measure the hazard from wall and ceiling linings, particularly when some new synthetic plastics are employed. Much additional work is required to determine to what extent and for which materials this test can adequately measure fire hazard.

Various organizations (e.g., Union Carbide Co., Monsanto Co., and the U.S. Forest Products Laboratory) have developed small tunnel tests as a more modest means for evaluating materials than the ASTM E 84 tunnel test. The designs of these tunnels allow flame spread ratings from the ASTM E 84 tunnel and the smaller tunnel to be correlated, but only for specific materials with flame spread classifications in a rather narrow range. Exact modeling between a smaller tunnel and the ASTM E 84 tunnel is almost impossible due to the complexities of the fire dynamics involved.

In the late 50's, the National Bureau of Standards developed the radiant panel test with the specific objective of providing a relatively simple and reproducible method for measuring the surface flammability of materials. It was expected that if such a test method could be provided it would be widely used for research and quality-control purposes during manufacture of building finish materials. This test method was recognized by ASTM in 1960 (E 162). It is seeing increased use in building codes and material specifications even though ASTM E 162 includes a statement that it is not intended for such purposes.

The test recognizes two important factors in characterizing surface flammability: (1) the critical energy flux necessary to propagate the flame and (2) the rate of heat liberated during flame spread. Moreover, the radiant-panel test method permits separate measurements of these two properties during a single test. The proportionality constant used in the calculation of flame spread index was selected to provide some measure of agreement between the resulting index and that derived from the ASTM E 84 tunnel method, particularly over the range of flame-spread indices of about 10 to 150. However, data with synthetic polymeric materials indicates that the correlation between these two tests is uncertain.

It seems fair to suggest that use of both the Steiner tunnel test and the radiant-panel method has contributed considerably to our understanding of the flammability behavior of solids. However, a need still exists to conduct carefully planned research programs on the relevance of surface flame spread ratings currently being used to determine fire hazard; particularly in the case of synthetic polymeric building finish materials. Prior to the completion of such programs it seems premature to assume that these or any other surface flame spread test methods are capable of rating materials correctly.

Some tests in which flame spread is a factor have been developed to measure a specific hazard. For example, DOC FF 3-71 flammability standard for children's sleepwear is intended to control the hazard of flame spread continuing upward if the fabric is ignited. Similarly, ASTM D 2859 (or DOC FF 1-70) measures the ease with which a small ignition source, simulating, for example, a burning ember from a fire place, could propagate a flame over a carpet or rug. These tests form the basis for flammability standards for children's sleepwear, rugs and carpets respectively. The standards are intended to prevent the occurrence of significant flame spread in the event ignition occurs. Both tests were developed in direct response to a specific hazard and are thought to be quite effective in dealing with that particular hazard.

Another test specifically designed to measure surface flammability of a specific item, i.e., roof coverings, in the configuration and under exposure conditions which might be typical of a real-life fire, is ASTM E 108. This test seems to be a reasonable approach to classifying roof coverings.

In summary, the surface flammability characteristics of materials are important and should be evaluated. However, since the flame spread process is strongly controlled by the magnitude and relative importance of the heat transfer modes involved, small-scale tests designed to provide relative comparisons of hazards have very limited meaning because they fail to model thermal energy feedback correctly.

On the other hand, small-scale tests developed to simulate a specific surface flammability hazard (e.g. carpet test) can be quite effective in controlling that particular hazard. At this time it is not clear what the most effective approach will be to classify the hazard from wall and ceiling finish materials. ASTM E 84 (or possibly ASTM E 162) and the corner test offer a potential method, without a complete full-scale burn, to evaulate these materials.

Much work must be done to relate surface flammability test ratings to full-scale fire behavior hazard, but it is unlikely that any one flame spread test can predict the surface flammability hazard of all materials in all situations.

5.4 Rate of Heat Release Tests

In recent years, there has been growing support among workers in the fire field of the concept that the rate at which heat is released during burning is an important criterion for evaluating the fire hazard from a particular material. It is considered to be a significant "characteristic" of room linings, and it must also be an important parameter of room contents that could contribute to early fire growth.

Since the rate of heat release is thought to be most important during the "steady" burning period following flame spread, it is a measure distinct from ignitability or surface flame spread potential, and is considered to be of greatest significance in the stage of fire growth preceding flashover.

The rate of heat release from initially ignited material(s) has a significant influence on local fire intensity and hence on the subsequent development of a fire since 1) the intensity of the fire strongly influences the probability of secondary ignition of nearby objects, 2) the fire intensity determines the rate of buildup of smoke and toxic gases in the room of fire origin as well as throughout the building containing the fire, and 3) all fire suppression techniques become ineffective once the fire intensity has grown beyond some critical level. Thus a basis for the importance of the rate of heat release concept is readily established.

The concept of rate of heat release had its origins in the early 1940's (Steiner, 1943) when the measurement of "fuel contributed" along with flame spread rate in the newly developed tunnel test (ASTM E 84) was described. While Steiner did not describe his plot of flue-gas temperature vs time as a rate of heat release curve, it was, in fact, a measure of the rate of heat release during the stages of surface flame spread and subsequent burning of the specimen. Thirteen years later, (Robertson, Gross, and Loftus, 1956) in what is now ASTM E 162, a flame spread rate was combined with a factor involving the rate of heat generated by the material under test to determine a flame spread index for that material. Both of these tests are essentially flame spread tests, and when evaluating heat release, these tests only report total heat released, and only provide information relative to red oak (ASTM E 84) or hardboard (ASTM E 162). However, reference to the time-gas temperature data could yield rate of heat release measurements. These measurements would include heat released during flame spread *and* steady state burning, in contrast to current proposed measurements which focus primarily on the steady burning period.

The first instrument designed specifically to determine rate of heat release was the FM Construction Materials Calorimeter (Thompson and Cousins, 1959). Recently, instruments have been developed at the National Bureau of Standards (NBS) (Parker and Long, 1972), Ohio State University (OSU) (Smith, 1972), Stanford Research Institute (SRI) (Amaro et al., 1974) and Forest Products Laboratory (FPL) (Brenden, 1975) to measure the rate at which heat is released from a burning material.

In a typical heat release rate calorimeter (HRRC), a sample of material, of known physical and chemical composition, is exposed to a controlled air flow and an external radiant heat flux simultaneously. In some instances, a pilot ignition source is provided at the bottom edge of the sample. The back surface of the specimen is either insulated or water-cooled. In the latter case, the temperature rise of the water stream is used to calculate the heat "released" through the specimen back surface. A secondary gas burner may be present in the flue stack.

Once the specimen is ignited, heat is released as a function of time. This released heat may be measured directly or by substitution, by either operating the calorimeter in an isothermal or adiabatic mode. In the substitution method, the flue gas time-temperature curve is reproduced by burning a gas, e.g. propane, to make up the difference between the test and reference (inert) samples. The rate of heat release of the test sample is thus obtained from the rate of consumption of propane during the substitution run. During adiabatic operation of the HRRC, the rate of heat release can be calculated directly from the rate of temperature rise in the products of combustion. In the isothermal mode, the flue gas temperature is monitored by thermocouples and kept constant by adjusting the energy input to the secondary gas burner in response to the unknown heat-release rate. The heat release rate is then calculated directly from the gas consumption rate. Table 5-2 summarizes some of the important features for the five calorimeters (Chamberlain, 1975).

The HRRC attempts to duplicate the path taken by materials in this consumption during a fire. Hence, the heat release rate measured with the HRRC is an extrinsic property of the test method which depends on a number of factors, such as: the specimen geometry and orientation, the availability of air and the air flow velocity, external heating, heat losses to the calorimeter walls, dripping, charring,

Table 5-2. Heat Release Rate Calorimeters

Location	Sample Size (cm x cm)	Type	Exposure	
			Source	Intensity (W/cm^2)
FM	122 x 122	Substitution adiabetic equiv. gas flow	Hot combustion gases	t/T curve, 0-12
NBS	11.4 x 15.2	Direct isothermal equiv. gas flow	Gas fired radiant panel	1.5-9
OSU	25.4 x 25.4	Direct adiabetic temperature rise	Electric radiant panel	Up to 3.5
SRI	46 x 21	Direct isothermal equiv. gas flow	Gas fired radiant panel	1.5-9
FPL	46 x 46	Substitution adiabetic equiv. gas flow	Gas fired panel	Up to 4

etc. Thus, since the rate of heat release is not a fundamental physicochemical property of a material, one would expect that different HRRC, employing different "fire conditions," would yield different measures of rate of heat release.

Each HRRC has one or more shortcomings. For example, both the Factory Mutual and Forest Products Laboratory HRRC's operate on the substitution principle and hence, they require two runs to evaluate each sample. The Factory Mutual calorimeter also employs an exposure on the sample which corresponds roughly to the standard ASTM E 119 time-temperature curve compressed into a much shorter time. However, the test suffers since it only allows one fire exposure. The primary disadvantages of the National Bureau of Standards instrument are the small sample size, the limitations in evaluating an assembly of materials in a practical configuration and the use of an after-burner. This latter feature may result in a higher value for rate of heat release than would occur in a real fire, where unburned pyrolysis gases might escape the room of fire origin. Both the Ohio State University and Forest Products Laboratory calorimeters are limited to the low and moderate heat flux ranges. The Ohio State University instrument also suffers in that the rate of heat release due to flame spreading over the surface cannot be uncoupled from the rate of heat release from those portions over which the flame spreading process is completed. This shortcoming could completely obliterate the initial heat release rate characteristic of, e.g., charring materials. Also, since the Ohio State University instrument operates adiabatically, with a direct measurement, the continuous application of a pilot flame at the specimen bottom edge will influence the heat release rates. The Stanford Research Institute instrument was specifically modeled on the National Bureau of Standards calorimeter, maintaining the high heat flux capability but increasing the size so that much larger specimens could be accommodated. Unfortunately, as shown in Table 5-3, the results obtained from these two instruments, supposedly similar in design, do not agree (Brenden, 1975). This difference is probably due to specimen size effect on the rate of heat release, but it is important to note the procedural differences between the two instruments such as top-versus-bottom piloting and back-surface insulation versus back-surface cooling. In fact, it appears that attempts to compare data derived with any of the five calorimeters will be unsuccessful in large measure because of the overriding dissimilarities in operation.

Table 5-3. Comparison of HHR Calorimeters

Douglas Fir Plywood Marine Grade	Heat Release Rate (W/cm^2)			
	NBS	SRI	OSU	FPL
Exposure (W/cm^2)	2.6	2.6	2.6	1.6-2.3
Time to Ignition (sec)	155	97	26	265
Peak HHR	8.8	10.8	10.5	17.0
Time to Peak (sec)	17.0	170	26	23 min
First 1-min. avg. HHR	0	0	3.2	0
First 5-min. avg. HHR	2.4	4.7	6.7	1.7

Current HRRC's are not restricted to any specific materials and thus are suitable for the determination of rate of heat release from polymeric materials. An example of an HRRC output curve is shown in Figure 5.1. This curve is typical and qualitatively correct for wood and other materials that form a stable char. The ignition delay, usually measured with a stopwatch, is the time from the closing of the door to the appearance of a flame. The peak heat release rate for these materials occurs at or shortly after ignition; the heat release then declines to a slowly diminishing plateau until the specimen is consumed. Other types of materials, e.g., those provided with a thin-layer of decorative coating, exhibit a much different response, e.g., high initial heat-release rate that rapidly falls to a much lower rate (Chamberlain, 1975). Thus, the technique has demonstrated the capability to detect and measure differences among various materials. From these curves, peak, one-minute average, five-minute average and ten-minute average rates of heat release are reported. Table 5-4 shows some HRR data for three manufactured wood products. However, a brief table of peak and average heat release rates cannot adequately describe the burning characteristics of a material. It should be accompanied by the HRR curve. For

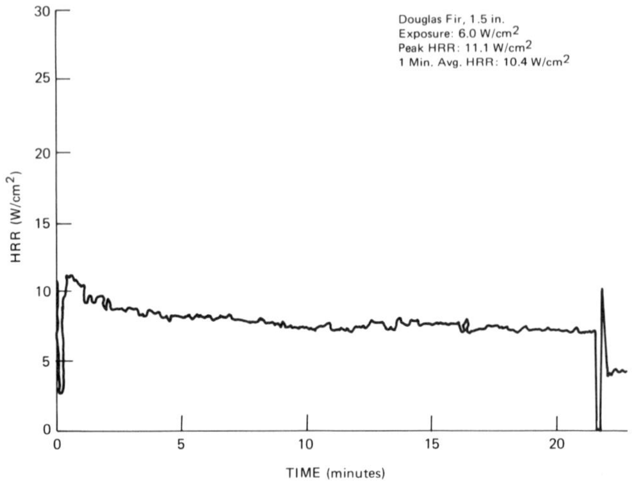

Figure 5.1. Typical HRR calorimeter output curve.

Table 5-4. Heat Release Rates for Wood Products Exposure = 6.0 W/cm²

Material (thickness mm)	Heat Release Rate (W/cm^2)			
	Peak	1-min. avg.	5-min. avg.	10-min avg.
Particle Board (16)	13.2	9.6	10.9	9.8
Hard Board Medium Density (10)	17.5	11.1	11.5	-
Hard Board Medium Density Paper-Faced (11)	24.9	16.2	11.6	-

example, the peak rate, 24.9 W/cm^2, for the hardboard, occurs as a sharp peak immediately after ignition, which quickly diminishes to a much lower and fairly constant value. On the other hand, the peak value for the unfaced medium density hardboard, 17.5 W/cm^2, did not occur until about seven minutes, and was followed by disintegration of the specimen. However, the average rate of heat release, and hence the total heat released, was almost identical for the two specimens during the first five minutes.

Table 5-5 presents data on southern pine from the NBS calorimeter which show reasonable reproducibility of HRR. However, the results from some preliminary round-robin testing on marine trade plywood shown in Table 5-3 indicates little if any agreement among the various HRRC. Explanations for the differences observed

Table 5-5. Performance of NBS Calorimeter Southern Pine at 6.0 W/cm² Exposure

Date	Peak HHR (W/cm^2)	Average HHR (W/cm^2)		
		1-min.	5-min.	10-min.
6/25/74	12.2	10.4	10.6	9.1
6/25/74	12.8	8.8	11.3	9.7
2/13/75	13.0	10.3	11.0	10.0
2/14/75	13.0	10.0	11.0	10.9

included the location of the pilot flame and basic design differences among the four instruments.

The development of a test method generally includes three stages: the growth of a concept, the design and development of an instrument to measure a quantity based on the concept, and the application of the measurements to a specific situation. Test methods for rate of heat release are presently at the second stage; consequently little if any data are available that correlates rate of heat release measurements with large scale fire behavior. This step should follow shortly and then criteria will be developed, based on rate of heat release tests, to improve the fire safety of materials in use. Meanwhile, more data are needed to fully evaluate the performance characteristics and reproducibility of the current available instruments. Current HRRC's differ significantly in their designs and capabilities, and most are still undergoing changes. Much more round-robin testing should be done to assess the accuracy and reproducibility of each instrument; therefore, persons involved with rate of heat release measurements should assemble and decide which, or a combination of which, instruments should form the basis for a test method.

A recently developed test, that in essence measures rate of heat release from wearing apparel, is known as the "Mushroom Apparel Flammability Test" *(MAFT). This test has been proposed (Brenden et al., 1976) as the basis for a new Federal Standard for flammability of wearing apparel.

Fabric specimens are tested in a cylindrical configuration (which simulates a garment) and pass-fail criteria are based on time to ignite them with a specified gas flame, and on heat transferred to sensors in the apparatus.

A fabric classification scheme, based on maximum heat transfer rate and minimum ignition time, is proposed as part of the standard, and suggests that fabrics which transfer little heat to the inside of the cylindrical configuration (Class I) could be used for all garments. Use of fabrics which transfer larger amounts of heat would be restricted as to garment type and style. The conditions of the test are intended to simulate realistic situations.

5.5 Tests for Oxygen Requirement

Since its introduction (Fenimore, 1966), the Oxygen Index Test has achieved wide popularity as a means of characterizing the flammability of materials. Its advantages include simplicity of equipment and ease of operation, good reproducibility, good differentiation of materials, versatility, and the use of small, easily prepared test specimens. These attributes make it a valuable tool for use in the preliminary stages of product development and for quality control purposes. The standard version of the Oxygen Index Test, ASTM D 2863, is applicable to plastics including solid, cellular, and film forms. A similar procedure has been used extensively for textile materials.

In the standard test, the sample is ignited at the top and burns downward with a small candle-like laminar flame. Energy from the flame is dissipated to the cool surroundings and there is little energy feed-back to support the combustion of the sample. This is in sharp contrast with the hot turbulent environment characteristic of most real fires. For these reasons, the test is of little value in characterizing the behavior of materials under conditions encountered in real fires.

Numerous variations of the standard test have been used for experimental purposes. Ignition of the sample at the bottom with upward burning gives a substantially lower value for the critical oxygen concentration than does the conventional top burning configuration (Stuetz et al., 1972). This lower value is due to more effective heating of the sample by the flame plume and, perhaps, better mixing of oxygen with the fuel gases. It has been found that a standard oxygen index of approximately 27 is needed to assure that a small sample of material will be self extinguishing when ignited at the bottom in air (20.9% oxygen).

Thermoplastic materials present problems in the standard test because of melting and dripping. The melt carries heat away from the combustion zone so a higher oxygen concentration is needed to maintain the heat balance at the burning surface. To avoid this problem, wicks or cup type sample holders have been used to hold the melt in place. Good correlations of oxygen index with polymer viscosity have been observed (Reimschuessel et al., 1973). This correlation may be important in studying the effects of additives on polymer flammability characteristics. Additives that lower the melt viscosity may increase the oxygen index, giving a false indication of flame retardant action which may not be corroborated by other tests.

The oxygen index decreases with an increase in the initial sample temperature since less energy feedback is needed to maintain the burning surface of the sample at the required temperature. This effect can be a cause of error when the apparatus becomes heated from successive tests.

Routley (Routley, 1973) has suggested making use of this effect by measuring the oxygen index as a function of temperature and taking the temperature at which OI = 20.9 (burning in air) as a measure of material flammability. This suggestion appears to be useful because almost all real fires involve pre-heated fuels at oxygen concentration at or below that of air. The oxygen sensitivity, the rate of change of OI with temperature in the atmospheric oxygen concentration region, may be another useful measure of flammability.

Numerous attempts have been made to provide a theoretical interpretation of the oxygen index test, but the results to date have been unsatisfactory (Martin, 1968; Johnson, 1974; Kanury, 1975). While the sample is burning, energy feedback from the flame to the burning surface maintains the surface temperature at a level where pyrolysis of the polymer supplies gaseous fuel to form a combustible mixture with the oxygen/nitrogen stream. As the oxygen concentration is decreased (and the nitrogen concentration increased) the flame temperature will decrease, reducing heat feedback and the supply of fuel to the flame zone. At a critical oxygen level (the oxygen index) a sudden transition from active burning to extinction occurs, but satisfactory understanding of this phenomenon requires both a detailed description of the energy transport processes in the experimental system and identification of the critical parameters leading to flame extinction.

Attempts have been made to apply the oxygen index test to liquids (Nelson and Webb, 1973). Indeed, the original concept for the test derived from studies of the burning of liquids and gases (Simmons and Wolfhard, 1957). From a practical standpoint, the standard test configuration suffers from the disadvantage that the sample holder provides a large heat sink with poorly defined heat transfer characteristics. The measured value of the oxygen index under these conditions is more a function of the design of the apparatus than an intrinsic property of the liquid. The study of liquids is attractive from a theoretical point of view since the vaporization process is simpler and better understood than pyrolysis of polymers. To avoid the heat transfer difficulty, Roberts (Roberts, 1975) devised a flow system where the bulk temperature of the liquid could be maintained at a constant value and heat loss to the environment was minimized. Using this method, he showed that the temperature at which the oxygen index is 20.9 is approximately equal to the fire point of liquid as measured by conventional methods such as ASTM D 92.

The difficulties encountered in these experiments with liquids help to show why the oxygen index test is successful with solid polymers. Most solid polymers have low and similar values of thermal diffusivity. The temperature of the burning surface is high and relatively constant for all organic polymers since it depends on the strength of the carbon-carbon bond rather than on the vaporization of small molecules. When an elongated polymer sample is burned at its upper end, heat loss from the hot burning surface into the sample and sample holder is small and relatively independent of minor variations in composition and sample size. The heat balance at the burning surface depends primarily on the chemical properties of the polymer and the composition of the atmosphere and is relatively independent of apparatus parameters. Thus the test provides a convenient way of comparing the relative flammability of similar polymer compositions under closely controlled laboratory conditions. Recourse must be had to test methods providing a much more severe fire exposure in order to predict the performance of materials under conditions likely to be encountered in real fires.

5.6. Extinguishment Test Methods, Specifications, and Standards

5.6.1 Portable Extinguishers

In terms of extinguishment, combustibles are classified by the National Fire Protection Association in four categories:

Class A: "ordinary combustible materials, such as wood cloth, paper, rubber *and many plastics"*

Class B: "flammable liquids, gases, and greases"

Class C: "energized electrical equipment"

Class D: "combustible metals, such as magnesium, titanium, zirconium, sodium, and potassium."

As is well known, well over half the tonnage of synthetic polymers in use consists of materials which fuse on application of heat to produce a low-viscosity melt. Polyethylene and polystyrene are major examples. On the other hand, some polymers produce a rather viscous melt which tends to stay in place rather than flow (e.g., polymethyl methacrylate), and yet others char rather than melt (e.g., phenol-formaldehyde).

Furthermore, different solids of the same size and shape which are fully ignited will burn at different rates. For example, Magee and Reitz (Maggee and Reitz, 1975) found that, for horizontal square slabs of 317 square cm area, the burning rate (gm/cm^2-sec) was three times as great for polystyrene as for polyethylene. Char-forming materials burn even slower relative to polystyrene, which is the fastest burning unfoamed solid of more than a dozen tested by Tewarson (Tewarson, 1975). On the other hand, liquid styrene monomer burns more than three times as fast as solid polystyrene, under certain conditions.

There is no standardized way of rating ease of extinguishment of various combustibles which takes into account these factors of different burning rate and different tendency to form a fluid melt, except in the sense of Class A versus Class B classification. As the quoted definitions suggest, plastics are generally taken to be Class A combustibles. Portable (hand-held) extinguishers intended for Class A fires are tested by approval agencies (UL, Factory Mutual) for their ability to extinguish a specified stack of hardwood pallets, after a specified preburn time. Also, excelsior (wood shavings) is sometimes used to represent a Class A combustible in these tests.

Class B combustibles are simulated by a square steel pan containing n-heptane. The larger the capacity of the extinguisher, the larger the area of test fire which must be extinguished in an approval test. For example, an experienced operator should be able to extinguish a 1.8 m^2 test fire with a 6 kg extinguisher containing dry chemical agent.

It is probable that a fire involving a polymer which forms a fluid melt can be controlled by an extinguisher intended for Class B fires. Application of water to such a fire could cause spattering and temporary augmentation of burning rate, but continued application of water would cool the polymer to a nonburning state. Note that since the heat of gasification of a molten polymer is several times as high as that of an organic liquid such as heptane or benzene, and also the flash point of the polymer is much higher, molten polymer fires would be considerably less difficult to extinguish than heptane fires. On the other hand, if the polymer is a smoldering cellular material, the fire being deep-seated, attack with a Class A extinguisher (rather than Class B) would appear to be more appropriate.

The foregoing paragraph is a summary of the fire protection engineer's knowledge about fighting polymer fires with hand-held extinguishers. There are no standardized test methods to determine the best extinguisher for a particular polymeric combustible, or conversely to determine the relative ease of extinguishment of a series of polymeric combustibles by a given extinguisher. Such test methods might be useful, but their value would be limited severely by two factors. First, a given portable extinguisher may be used to fight fires of a wide variety of materials, making optimization difficult. Second, the difficulty of fighting a fire depends more on the size of the fire than on the type of combustible, so factors such as spread rate, time of detection, ready availability of extinguisher, etc., may be much more important than specific effectiveness of a given extinguishing agent on a given polymeric fire. Hence, the need for better test methods in this area would appear to have low priority.

5.6.2 Automatic Sprinklers

Automatic sprinklers to protect buildings are specified on the basis of the delivered water density (generally between 0.1 and 0.6 gallons/min-square foot) and the maximum foreseeable water demand (sometimes as large as several thousand gallons/min). The cost of installing the sprinkler system is highly dependent on these design parameters.

The water density required to control a fire obviously depends on the type, quantity, and arrangement of combustibles present (Magee and Reitz, 1975). Polymeric materials in some cases, because of high burning rates, may present especially severe challenges to sprinkler systems. However, in general, fire loss to property, especially commercial and industrial property, has been tremendously reduced when automatic sprinklers have been present. Insurance statistics show that the expected loss due to fire in sprinklered properties is about $2 per year per $10,000 value. Data suggest that it is about an order of magnitude higher for unsprinklered property.

Private residences are virtually all unsprinklered, because of cost. A minor proportion of hotels, school dormitories, etc. are sprinklered. A greater proportion of nursing homes, restaurants, etc. are sprinklered, depending on local ordinances. An accurate knowledge of the minimum water application rate and quantity required to control fires of polymeric materials would be of great value in judging cost effectiveness of introducing sprinklers versus modifying or banning polymeric combustibles to achieve a desired level of fire safety.

Standards for acceptable sprinkler protection of commercial and industrial properties have been obtained and are continually being refined by insurance companies (e.g., Factory Mutual), partially by full-scale fire tests and partially from loss experience. In a few cases, modeling experiments have been helpful. Much more information is available on industrial and commercial than on residential combustibles.

To illustrate a difference in burning of wood versus expanded polystyrene under sprinklers, the following data from Factory Mutual are quoted:

Common Conditions: Pallets made of either wood or expanded polystyrene, stacked six feet high (9 stacks in square arrangement), with sprinkler density of 0.3 gal/min-square foot, top of array 9 feet below ceiling, 165°F sprinkler links spaced 10′ by 10′ on ceiling.

Results – Wood: Seven sprinklers opened; maximum ceiling temperature above fire was 565°F. The fire spread was controlled.

Results – Polystyrene: 59 sprinklers opened; maximum ceiling temperature above fire was 2100°F. Fire burned intensely until melting caused collapse of the pile.

For very intense fires, especially under high ceilings, the upward velocity of the fire plume prevents penetration of water droplets. By modifying sprinkler design to increase production of large drop sizes, penetration can be dramatically improved. By operating sprinklers above a fire plume maintained and closely controlled by feeding gaseous fuel to a ring of burners, and measuring water arrival at floor level in the center of the ring, one can measure penetration efficiency as a function of sprinkler parameters (it is also desirable to produce some fine drops for optimum cooling of the plume and ceiling jet, to prevent damage to the ceiling and limit excessive opening of sprinklers.) Factory Mutual is currently attempting to develop sprinkler standards based on these considerations.

One of the most severe fire challenges known, other than flammable liquids or gases, is that of plastic commodities which are stored in factories, warehouses or retail outlets, in configurations reaching to substantial heights. The maximum tolerable height under ceiling sprinklers has been empirically determined (Dean, 1975) for various commodities, and must be much more restricted for plastics than for cellulosic materials. Among various plastics, Dean found trends as follows:

polystyrene – vigourous burning
polyethylene, polypropylene – intermediate
polyvinyl chloride – relatively mild

However, he noted that exceptions to such rankings commonly exist because of effects of detailed geometry, packaging material, or additives, especially plasticizers.

It may be remarked that it is possible to store polymeric materials safely to greater heights than those permitted under ceiling sprinklers by installing additional sprinklers at intermediate levels, preferably below horizontal barriers. Alternatively, a development program is underway at Factory Mutual to use unconventional sprinkler heads of greater penetrating power at ceiling level, which would permit relaxation of either storage height requirement or total water demand.

In contrast to industrial fires under high ceilings, there is very little experience bearing on the minimum sprinkler requirement for a fire in a normal-size residential room. There is reason to believe that very little water may suffice to control such a fire. However, if it is necessary to spray the water directly on the burning object, there is difficulty in locating a sprinkler head so that the spray will impinge on every potential combustible in the room; if several sprinkler heads are needed per room the installation cost will be high. There has been developed data relevant to another concept – i.e. spraying the sprinkler water into the fire gases and onto the hot ceiling and walls so as to generate steam which may extinguish the fire by inerting (Kung, 1975). Less than 0.05 gal/min-square foot may be adequate. Much testing is needed with realistically furnished rooms before standards can be specified.

The action of the sprinkler to control the fire is in part by inerting (both by steam and by entrained combustion products), in part by direct wetting of the pyrolyzing solid, in part by pre-wetting not-yet-ignited materials in part by cooling the fire gases and thus reducing radiative transfer to combustibles, and in part by creating a fog which itself blocks radiative heat transfer. In view of these multiple effects, there seems to be little chance that a small scale test method evaluating any one of these effects will suffice to determine minimum water application rate for a given polymeric commodity.

5.7 Tests for Smoke Evolution

Tests for smoke evolution are important to fire safety because the presence of smoke affects the ability of occupants to escape from a burning structure and the ability of firefighters to control and extinguish the fire. Most existing smoke methods are concerned with vision obscuration, and therefore seek to measure smoke density by either optical or gravimetric techniques. No smoke method presently in use addresses an important aspect, the lachrymatory and irritant characteristics of smoke which can obscure vision even more effectively than optical density in the atmosphere.

The National Bureau of Standards smoke density chamber is a widely used apparatus for measuring smoke density. This chamber represents a means for evaluating the relative smoke-producing characteristics of a material in terms of obscuration, and much of the discussion here addresses it because there is only one other test method which could be considered a serious contender, the ASTM D 2843 test.

The NBS smoke density test method, also designated NFPA 258, exposes a specimen 3 inches square to a radiant heat flux of 2.5 W/cm^2, either in the presence or in the absence of a pilot flame (flaming or smoldering mode, respectively). The smoke evolved is accumulated in an enclosure with a volume of 18 ft^3, and optical density is measured along a vertical light path 3 ft long. The results are expressed in values of specific optical density (D), and the maximum value is designated as D.

It is when the use of the test method is expanded from its basic function of obtaining comparable data on materials that some aspects can become questionable. As examples, consider the following:

1. The maximum value of specific optical density, D, is corrected by subtracting the optical density due to soot deposited on the optical system. This procedure favors heavy soot-depositing materials such as polystyrene to the possible detriment of fire safety. The soot deposited on the exposed surfaces of the optical system was obviously in the atmosphere obstructing vision at some time during the test, and its propensity for depositing on an exposed surface makes it even more likely to obscure vision through irritation and lachrymation, yet the correction procedure rewards rather than penalizes this characteristic behavior. The Federal Aviation Administration proposal to use D values at 1.5 and 4.0 minutes is to be commended because it avoids the correction step. Fortunately for public safety, the D values for heavy soot-depositing materials are so high, even with the correction, that these materials are still recognized as hazardous.

2. Specific optical density is a useful scale for expressing relative smoke density. Its use in extrapolating values to larger enclosures, however, can be questioned. As smoke density increases, multiple scattering effects may become appreciable. It is generally agreed that multiple scattering is not significant at D values up to 16, but extrapolating D values over 200 to larger enclosures is of doubtful significance.

3. The vertical specimen position permits materials which melt to exhibit unrealistically low values because the molten material either is subject to less heat in the specimen holder trough or overflows the trough and escapes exposure entirely.

The NBS chamber offers the advantages of considerable versatility, such as the addition of controlled ventilation, continuous weight monitoring, and analytical and bioassay devices for measuring toxicity. It offers one-dimensional heat flux, which is essential for evaluating composite structures which are expected to encounter heat flux on only one side.

The ASTM D 2843 test is a useful laboratory screening test, and can distinguish gross differences between materials at substantially less cost than using the NBS smoke chamber. The horizontal photometer light path at a fixed height tends to favor materials which produce dense smoke that may not rise to the level of the light path.

The ASTM E 84 test is primarily a test for surface flame spread, however in the past a great many materials have been tested for smoke evolution by this test method. It is desirable that smoke evolution be measured in a test specifically designed for this measurement.

5.8 Tests for the Assessment of the Toxicological Aspects of Pyrolysis and Combustion Products

In general, smoke and toxic gases are far more important than heat and flame as a cause of fatalities in building fires. According to our present knowledge, carbon monoxide is the chief toxicant. However, with the introduction of new synthetic polymeric materials and products into building construction and furnishings, the potential for the production of new toxicants, e.g., HCN or HCl now exists. The situation is further complicated by the use of fire retardants to inhibit ignitability and surface flame spread. During combustion, certain retardants under some conditions can combine with elements in the polymer to form extremely toxic airborne products. The French, in an attempt to reduce this hazard, have promulgated a "toxicity standard," which simply sets an upper limit to the percentage of various elements, e.g., Cl, permissible in a given polymeric formulation. This approach is much too simplistic to insure control of toxic products.

When a polymeric material is thermally decomposed, under either oxidative pyrolysis or flaming combustion conditions, a mixture of substances comprising gases, aerosols, particulates and chars is produced. It is difficult to establish whether the intoxication syndrome (the signs and symptoms of intoxication produced on exposure to these decomposition products) is produced as a consequence of the action of one, some, or all of the substances present in these products. Furthermore, combustion toxicology is inherently complex because relatively minor variations in testing procedure frequently result in major changes in the actual dose of toxicants produced. Such testing variations include the level of incident heat flux used, the duration of heating, whether combustion is flaming or nonflaming, the physical configuration of fuel, the relative mass of the material degraded, amount of air flow used, etc. Essentially these variations represent a sampling of the various parameters which are constantly changing in a real fire.

The purpose for investigating the toxicity of a material's pyrolysis and combustion products is to permit the evaluation of potential hazards that might arise from the use of that material in a product or system. Fire safety toxicity standards for materials have largely ignored the complexity of the problem and are many times grossly over-simplified. A statement such as "shall be no more toxic than wood," used for example in some building codes, is virtually meaningless. It does not indicate how this determination should be made, nor whether the proposed material is used in comparable amounts or under equivalent conditions. Further, the toxicity of the decomposition products of wood varies greatly depending on species composition and combustion conditions. The present situation is confused and may even be described as chaotic.

Approaches to Toxicity Assessment

Tests for toxic gas assessment use one, or both, of two approaches: 1) identification and analysis of the chemical compounds in the decomposition products, and 2) exposure of laboratory animals to the decomposition products; in sum, analytical or bioassay.

Analytical

The analytical approach is based on the philosophy that if only a limited number of chemical compounds are responsible for essentially all the toxic effects, then analysis for these particular compounds would permit prediction of essentially all toxic effects which could be expected. However, the analytical approach suffers for the following reasons. First, a basic weakness of the analytical approach is the assumption that knowledge of the complex mixture of decomposition products produced under a given set of fire conditions can lead to satisfactory prediction of their combined toxic effects. Second, the biological activity of a compound may make it of great importance in the production of the intoxication syndrome, but its concentration may be very low and consequently not suspected to be significant, or it may not even be detectable by "common" techniques.

For these reasons, it should be mandatory to use laboratory animals during the first-tier screening assessment of combustion product toxicology.

Bioassay

Whereas the analytical approach calls for analysis followed by prediction of toxic effects on the basis of incomplete data for pure gases, the bioassay technique directly measures the effect of the combustion products on a biological system, and bypasses all the intervening uncertainties. Since human beings cannot be used in such experiments, a choice of test animal must be made, and mice and rats are the species of choice in most tests. There are differences of opinion as to which species, and which strain of that species, is more suitable for predicting human response. Rats are the preferred species for general toxicologic studies since they provide a more ready supply of blood for analysis. Mice are recommended for sensory irritation studies and some behavioral tests.

Apparatus

Although many apparatuses and procedures have been developed, there is no standard and/or procedure utilizing the bioassay aproach which is widely accepted. The available test apparatuses differ in size, location of the test animals, method of toxic product production, and amount of air flow used. The animal exposure chambers range in volume from 2.2 liters to 510 liters. The smaller the chamber, the greater the effect of oxygen consumption by the test animals over a period of time (unless pure oxygen is introduced to maintain the oxygen level at 21 percent), but small chambers limit the number and size of the test animals. On the other hand, the larger the chamber, the greater the quantity of material required to achieve a given concentration of toxicant, and the greater the probability of variations in gas and aerosol composition between different locations in the chamber. The exposure apparatus must be of sufficient size so as not to inhibit the animals' normal biological functions during the time of exposure.

It is necessary to minimize heat stress on the animals during a test. The available tests differ as to whether the test animals are in the same chamber as the material being tested. The major reason for separating the combustion chamber by a given distance from the animal chamber is to cool the gases sufficiently and minimize direct radiation so that the animals are not exposed to extreme temperatures. Theroretically, having a sample and animals in the same chamber would minimize loss of toxicant; in practice, the animals are kept at least 6 to 12 inches away from the sample to minimize heat stress. The time of travel of the combustion products and the accompanying heat losses can drastically affect the composition of the toxic species to which the animals are eventually exposed. Furthermore, reactions with and/or condensations on the exposed surfaces of the apparatus may also affect exposure conditions.

Toxicity tests also differ as to whether a fixed temperature or rising temperature history is used to generate the combustion products. Many materials produce more toxic effluents at some temperatures than at others, and the selection of fixed temperature levels unavoidably favors some materials over others.

There are differences between tests as to amount of air flow used. Some tests are entirely closed-system tests, with neither inflow of fresh air nor bleeding off of toxicants by displacement. The results are independent of flow rate and involve no loss of toxicants, but the lack of a fresh air supply causes oxygen depletion. Some tests use closed systems but add continuous air recirculation. Other tests use the flow-through technique with a continuous stream of air; this technique offers the lowest oxygen depletion and the highest level of oxidation during pyrolysis. Since oxygen depletion occurs in most real fires, the range of oxygen levels that can be considered realistic or allowable for the test animals becomes debatable.

Most toxicity tests use pyrolysis, and a minority use flaming combustion. Flaming combustion introduces more variables such as air flow rate, area of burning surface, etc., and aggravates the problems of oxygen depletion and heat production as regards the test animals.

Procedure

An appropriate screening test procedure should incorporate both a behavioral end point (loss of avoidance or other means of evaluating the animal's incapacitation) and a bioassay (appropriate monitoring of the physiological status of the animal such as blood chemistry, heart rate, breathing rate, etc.). Time of death is frequently taken as an end point because of convenience, but incapacitation may bear a more meaningful relation to fire hazard.

It should be obvious that the use of only one set of combustion conditions to evaluate a material or to attempt to compare different materials at a single condition may be inadequate. There is no *a priori* reason to expect that the behavior of a material under any one set of conditions is representative of its behavior under other sets of conditions.

Thus, a matrix must be used to provide an appropriate set of exposure conditions which may then be used to assess a material's potential toxicological hazard. The material must be exposed to nonflaming as well as flaming conditions to permit evaluation of its contribution to hazard in the smoldering state, at the point of flaming ignition, and during the post-flaming state. In addition to varying the heat flux and nature of combustion conditions, the matrix must also provide data pertaining to concentration-response or dose-response (a dose indicates the amount of chemical administered at a specific interval of time). The simplest procedure is to select a sample weight, perform the test, count the number of dead animals at the end, and report percent mortaility. This procedure provides one scale of comparison, from 0 to 100 percent; however, when a wide range of materials are tested at a single exposure level, many tend to fall either at 0 or 100 percent. The single result of either 100 percent death or 100 percent survival tells relatively little about the lowest dose likely to be lethal and how the material under consideration compares with other materials. Preferably, comparison should be made on the basis of response at several dose levels, as shown in Fig. 5-2, which illustrates the type of graph resulting from doses on a log scale plotted against percent mortality for two substances identified as A and B. From the graph it is clear that at dose X, both A and B are equally lethal, whereas at dose Y, A is more lethal than B, and at dose Z, B is more lethal than A. No single point would give this information. Incapacitation data can be treated similarly.

Summary

In a typical uncontrolled fire, conditions are continually changing, resulting in varying exposure conditions, e.g., heat flux, air flow, etc. Hence, the toxic environment is continually changing as well. Small scale tests are designed to give a measure of the specific toxicity (toxicity per unit mass consumed) of a material. In a real fire the toxic hazard will depend on the concentration of combustion products in the atmosphere, i.e, on the quantity of material burned, the rate at which it burns, the volume of the system in which the products are dispersed, as well as on the specific toxicity of the material. Thus, it is erroneous to expect that simple toxicity tests offer a precise hazard evaluation. Rather, well developed small-scale tests may be profitably used to screen comparable candidate materials under specified degradation conditions and permit the recognition of toxicologically unique pyrolysis or combustion products.

5.9 Tests for Fire Endurance

Fires usually start in buildings with one small item in flames, such as a waste paper basket or chair, and then grow in size. If the fire is going to become serious, the small fire which began with a single item eventually grows to involve the whole room. The instant of total room involvement is called "flashover."

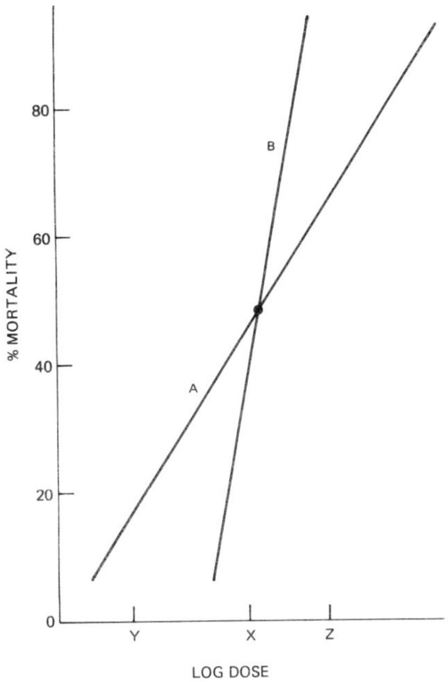

Figure 5.2. Comparison of mortality produced at dose levels X, Y, and Z by substances A and B.

The chain of events that lead to a serious fire can be broken in many places. Often the probability of a fire reaching flashover is very small, but a sequence of improbable events can lead to flashover even in the best protected areas. The realization that the worst eventuality is not impossible must lead to an evaluation of how a building would respond to a sustained high-intensity fire. This fire performance of a structure is termed fire resistance or fire endurance and is measured in the ASTM E 119 Fire Test.

One of the most important characteristics of the post-flashover fire is that it can be considered a volume process where *average* temperatures and heat fluxes within the compartment are meaningful concepts. This phase is directly contrasted with the pre-flashover period, during which time flames are either localized to stationary sources or else characterized by flame fronts advancing along surfaces. The gas temperatures have extreme spatial variations — for example, flame temperature near 2000°C in some areas and near-ambient temperature elsewhere.

A number of factors, such as ventilation, fuel load, and thermal properties of the structure, have a controlling influence on fire intensity at any instant. To illustrate such influence, the historical development of a standard fire exposure is reviewed and its relationship to the expected fire is discussed: The earliest attempts to test the fire resistance of different structures occurred in the last two decades of the 19th century and essentially consisted of some *ad hoc* furnace tests in which the specimen was exposed to a constant temperature for a given time period. Sixty years ago, an ASTM committee issued the first edition of C 19, the predecessor of the current E 119 standard (ASTM, 1918. The C 19 standard was based on an exposure of the test specimen to conditions based on a time-temperature curve which was believed to approximate the course of fire in the heavily timbered structures of that day. This curve has been essentially unchanged and to this day it closely resembles the curves used by most countries for the post-flashover evaluation of structures. The ASTM E 119 curve and those of other countries are shown in Figure 5.3. It is quite evident that they have common origins. The concept of using more than one exposure was never adopted and these curves have become the only post-flashover exposures for structures used in buildings (ASTM, 1918).

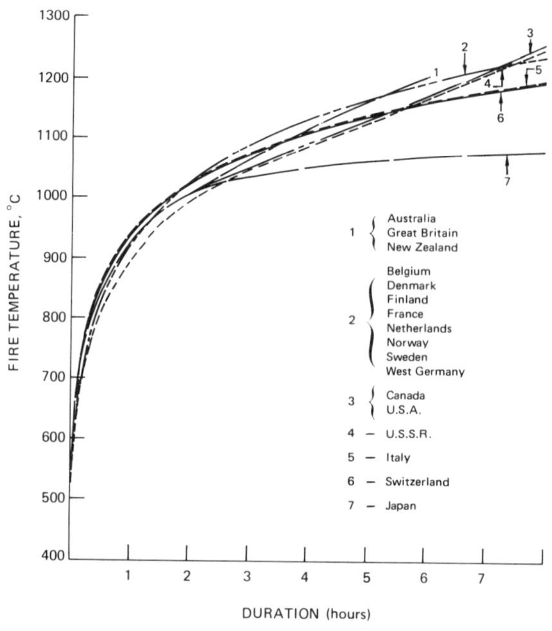

Figure 5.3. Standard time-temperature curves for fire tests of building construction and materials (post-flashover).

After the initial adoption of the standard post-flashover fire test, some interest arose in checking the validity of the assumptions that underlay the committee's work. The most significant research was attempted by Simon Ingberg at the U.S. National Bureau of Standards (Ingberg, 1927, 1928, 1942, 1957, 1967). Starting in 1922 and continuing through the 1940's he conducted numerous research programs aimed at better characterizing the post-flashover fire.

Ingberg's experimental work probed two aspects of describing the post-flash-over fire — conducting burnout tests and making fuel load surveys. The burnout experiments were conducted over a period of years in three different test buildings and encompassing both residential and office occupancies. An average time-temperature curve for one of these tests is shown in Figure 5.4 for a case where the fire was much less intense than in the standard exposure. Other experimental fires generated temperatures substantially greater than the standard exposure. One aspect of these tests was that Ingberg attempted to obtain the worst possible fire conditions by controlling ventilation. He normally adjusted swing shutters on the windows of his burnout chambers to achieve his purpose, but unfortunately never reported the ventilation used in his experiments. The systematic study of the effects of ventilation was not started until after World War II.

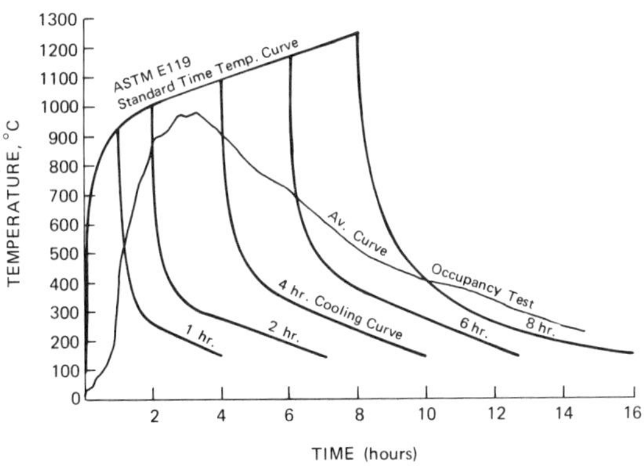

Figure 5.4. Average time-temperature curves for burnout tests (after Ingberg).

From the earliest tests it was apparent that these fires did not reproduce the ASTM time-temperature curve very well. A single standardized curve was so appealing, however, that Ingberg devised a way of molding reality to fit the curve. The strategem that he invented was to define a fire "severity" which was set equal to the integral under the time-temperature curve, above a baseline of either 150°C or 300°C. As can readily be proved, this equal-area "severity" has no physical justification. It did, however, conveniently reduce the fire description from a two-variable into a single-variable problem, so that different fires could be directly compared. The standard curve was thus saved — any actual fire could be defined to have an equivalent duration on the standard test, as shown in Figure 5.5. The final results were expressed in the following table:

Combustible Content lbs/ft^2 of Floor Area	Equivalent*BTU/ft^2	Standard Fire Duration (hrs)
10	80,000	1
15	120,000	1½
20	160,000	2
30	240,000	3
40	320,000	4½
50	380,000	6
60	432,000	7½

*Cellulosic materials have heats of combustion around 8,000 BTU/lb. However, many of the new synthetic polymers have heats of combustion much higher, 12,000–16,000 BTU/lb.

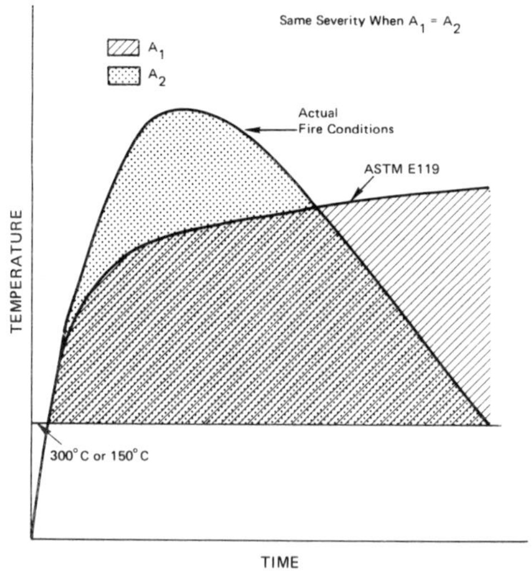

Figure 5.5. Fire severity equivalence time-temperature curve (after Ingberg).

To make these results useful, the expected fuel load has to be known. Fuel load surveys were conducted by NBS (Ingberg, 1957; Bryson and Gross, 1967; Culber and Kushner, 1967; Culver, 1976) at several times, and are still continuing today. They were intended to provide a rational basis for code classifications of occupancies, but in practice they have not been reflected in many codes.

The most important aspect of the ASTM E 119 fire test, and all similar post-flashover tests, is that it purports to allow the evaluation of performance of a structural element under a standard fire exposure. It is not used to evaluate whether the structural element will contribute to the propagation of a fire within a compartment, but rather the load carrying capacity or the fire containment performance of the element.

If one neglects the lack of realism in the standard time-temperature curve and Ingberg's definition of severity, then the ASTM E 119 test can be used to compare various building structural components. Polymeric materials have been used in many assemblies that have passed the ASTM E 119 test for period of time from twenty minutes to two and even three hours. In most cases the polymeric material has been protected by a thermal barrier, such as gypsum wall board, and the polymers were not involved until the later portion of the test period.

Maintenance of a positive pressure within the furnace would probably improve the ASTM E 119 test method, but at this time there are no requirements for measuring and/or regulating the relative pressure between the inside and outside of the furnace. Some laboratories conduct the test with a negative pressure over the entire face of the specimen which results in room temperature air being drawn through any openings in the unexposed surface of a wall or floor specimen. This means any flaming within the assembly would have a tendency to be drawn into the furnace rather than out the opening. Under actual fire conditions in fully involved compartments, a positive pressure exists in the upper two thirds of the compartment. It is particularly important that E 119 tests be conducted with a positive pressure on at least the upper two thirds of wall specimens if there are openings in the wall. One example of where this is important is when pipe and conduit penetrations, are being tested in the ASTM E 119 test. Another case is when doors are tested by ASTM E 152 which is the special version of ASTM E 119 that deals with doors. As with pipe penetrations it is important to conduct the test with a positive pressure at the top of the door to simulate actual fire conditions.

5.10 Full-Scale Fire Tests

Full-scale fire experiments have been the traditional precursors of standardized fire tests. Some of the earliest experiments were conducted in the 1790's (Hamilton, 1959) and there were extensive efforts in the late 1890's to develop quantitative measures of post-flashover fire performance. These led to the standardization of the fire endurance tests such as the ASTM E 119 discussed in the preceding section. There were subsequent full-scale experiments conducted by Ingberg in the 1920's and 30's at the National Bureau of Standards to further establish the meaning of fire endurance.

More recently, there has been increased activity in the use of full-scale tests, particularly to evaluate the contribution of synthetic polymeric materials to early fire growth. In most cases, these tests have been *ad hoc* evaluations of fire performance which have not been based on any standardized procedure.

The need for large-scale tests has grown as many people in the fire safety community come to realize that the existing small-scale tests are not adequate to evaluate fire hazard in a consistent manner for all materials. The discrepancy between simulated real-life situations and small-scale tests has been well documented. The problem was brought to a head when the Federal Trade Commission issued their proposed rule on cellular plastics products (Federal Trade Commission, 1975). The Commission said, in effect, that the "Standard Method of Test for Surface Burning Characteristics of Building Materials," variously denoted as ASTM E 84, UL 723, NFPA 225, did not, alone, constitute a reasonable basis for certification, since it did not determine the combustion characteristics of the materials under actual fire conditions.

As a result of the FTC action there has been an intensified search within both industry and the building code community to determine what is a reasonable basis for substantiating performance of materials. The end result of this search has more or less centered on the concept of using a room test for the evaluation of wall and ceiling linings in lieu of, or as a supplement to, the existing ASTM E 84 test procedure. ASTM Committee E-39, in response to the need for a standard room test established a Task Group to develop a recommended practice. As a result, Task Group 4 of ASTM E-39.10.01 has developed a Guide for Room Fire Experiments (ASTM, 1977), to provide direction to the research or testing agency. The purpose of the Guide is to provide a document which compiles the information and experiences now available from those actively engaged in this type of testing. The Task Group hesitated at this time to come up with a test method, since the state of the art is still in development. However, the document that was produced does attempt to give in detail the various factors which should be taken into account in the design of a room or compartment test.

The Task Group discusses in their Guide such things as room size and shape, ventilation, specimen description, ignition source, instrumentation, and safety considerations. These represent an important range of choices which must reflect to some degree the nature and type of fire scenario to be protected against. For example, just looking at the decision to be made on the size, nature and type of the ignition source, indicates the complexity involved in deciding on the nature of a room test.

The ignition source in any fire will have a large effect on the total performance in the room. In fact, the ignition source may become a critical factor in determining how and when a lining may be involved in a fire. In some of the earliest work on evaluating linings (Corson and Lucas, 1950), an extensive study of ignition sources was done, preparatory to conducting a series of full-scale room tests to evaluate linings and wall constructions. The tests were conducted in a 14 × 20 ft. room 12 ft. high. A series of wood cribs were used of 5, 7½, 10, 20 and 30 lbs. weight. The heat release of the fire was analyzed; and the temperatures developed in the room and the maximum flame height were recorded. For a room of this height it took a 20 lb. crib to have the flames reach the ceiling.

Whether the flames reach the ceiling during a test will determine the relative importance of the combustibility of the wall and ceiling linings. If the flames do not touch the ceiling then the combustibility of the wall linings determines if the ceiling material will become involved, since the fire can only reach the ceiling by traveling up the wall lining. If, however, the flames reach the ceiling because of the size of the ignition source then the combustibility of the wall linings will be a lesser or negligible factor in determining if ignition and spread occurs on the ceiling. To illustrate this point (Fang, 1975), it was found that the importance of the wall relative to the ceiling lining depended upon the size of the igniting fire. For an igniting fire that did not touch the ceiling – simulating a small upholstered chair – the combustibility of the wall lining became a major factor in the time to flashover in the room. However, when a larger simulated upholstered chair was used, flames impinged directly on the ceiling and the effect of the wall lining became far less important; the combustibility of the ceiling lining governed the time to flashover. In this case, the choice of the size of chair can lead one to differing conclusions on the importance of the wall versus the ceiling lining.

The room test concept has also been adopted by the International Conference of Building Officials. In Section 1717c, in a recent amendment of their building code, they state that plastic foam may be specifically approved, based on approved diversified tests such as, but not limited to, tunnel tests, fire tests related to actual end use such as a corner test, and an ignition temperature test. As a result of this change, the Research Committee of the Uniform Building Code has been attempting to develop a standardized corner test to be used in conjunction with the application of their code.

A room test can serve many different purposes; two common objectives are:

1. As an acceptance test to qualify products or constructions for use as wall and ceiling linings.
2. To establish the validity of various small-scale tests, which in turn may be used in lieu of the large-scale room test as acceptance tests.

Obviously, the desired objective would be (2), but until suitable small-scale tests can be found which can be correlated to large-scale fire behavior, room tests may be the only means to adequately evaluate the hazard from interior finish materials.

It has long been suggested by many workers in fire research that the corner formed by the intersection of a ceiling and two walls represents a critical configuration for evaluating the flammability of interior finish materials. For example, a "corner-wall" fire test was one of the comparative tests used (Forest Products Laboratory, 1962) to relate the 8-foot tunnel furnace to realistic fire situations. More recently, the Factory Mutual Research Laboratory has reported (Maroni, 1972) a large-scale corner test to represent an industrial situation, and Williamson reported on a corner fire test to simulate residential fires (Williamson, 1973). Results from the latter test indicate that some polymeric materials, e.g., polyurethane foam, which obtained low flame spread ratings in the 25-foot tunnel Test (ASTM E 84), exhibited intense combustion and high fire hazard when tested in the corner test. Williamson concludes that the corner test is a meaningful test procedure for evaluating the hazard of the newer polymeric building materials. He now reports that the University of California, Berkeley can run this test for the same cost as the 25-foot tunnel test (Williamson, 1976).

It would seem that, until better test methods are developed which more accurately measure the fire hazard from interior finish materials, large-scale fire tests, e.g., room tests and corner tests, will be used to evaluate materials and products for acceptability.

5.11 Conclusions and Recommendations

Conclusion: For a long time, test methods have been developed with limited understanding of the phenomena of fire growth. Knowledge has advanced considerably in the last decade, and the development of more meaningful test methods based on fire dynamics can now be attempted. *Recommendation:* Require that tests and standards demonstrate a rational relationship to the hazards they are designed to control.

Conclusions: Many test methods and criteria for fire hazard used today were developed through experience with cellulosic materials. These tests are not necessarily adequate for the evaluation of fire hazard of synthetic polymers. No single test is adequate to completely evaluate the fire hazard of a particular material. The complexity of *ignition* phenomena makes it impossible to devise a single test for the evaluation of ease of ignition which is applicable to different materials and end-use situations. *Recommendation:* In the development of tests for *ignitability,* simulate well-defined ignition sources and end-use conditions.

Conclusion: The rate of *flame spread* is strongly dependent on many physical and geometrical parameters. Thus, tests which presume to establish relative flame spread characteristics of polymeric materials in a definitive manner may be valid only for the particular conditions of the test and thus may give misleading results if extrapolated to other conditions. *Recommendations:* Continue work on tests for fire hazard of interior finish materials; the currently used ASTM E 84 test must be improved or replaced. Correlate *surface flammability* test ratings to full-scale fire hazard.

Conclusion: The rate at which *heat is released* during burning is an important criterion for evaluating fire hazard from a particular material. Therefore, tests measuring rate of heat release can be of great significance. *Recommendation:* Continue the current development of tests for *rate of heat release.*

Conclusions: The *oxygen index* test provides a laboratory technique for the evaluation and guidance of the development of new materials; however, since it does not model energy feedback realistically, it is not sufficient for characterizing fire hazard, or for standards or regulations. There are no standardized test methods for determining relative *ease of extinguishment* for different polymeric materials. In view of the many factors influencing extinguishment, there seems to be little chance that a small-scale test method could be developed to adequately determine minimum water application rate for a given polymeric material. Tests for *smoke evolution* are important, since the presence of smoke is an important parameter of fire hazard. Available tests for smoke evolution yield results which depend greatly on the conditions of testing (e.g., ventilation, heat flux level, sample orientation, etc.). *Recommendation:* Place high priority on the development of more meaningful tests for *smoke evolution.*

Conclusion: Fire safety *toxicity standards* for materials have largely ignored the complexity of the problem. No accepted toxicity tests are currently available. *Recommendations:* Mandate the use of laboratory animals during a first-tier screening assessment of combustion product *toxicology.* Incorporate both a behavioral end-point and a mortality count in an appropriate *toxicity* screening test.

Conclusion: Currently available tests for *fire endurance* are generally satisfactory. *Recommendation:* Run the ASTM E-119 *fire endurance* test with a positive pressure on at least the upper two thirds of wall specimens so as to improve the test when there are openings in the wall.

Conclusion: Large-scale tests are necessary as acceptance tests for polymeric materials/products and to establish the validity of various small-scale tests. *Recommendation:* Reflect the nature and type of fire scenario to be protected against when designing *large-scale* room tests.

CHAPTER 6

TEST DEVELOPMENT BASED ON FIRE DYNAMICS

6.1 Introduction

If a material were expected to be exposed to fire risk in only one precisely defined set of circumstances (size, orientation, type of ignition source, method of applying ignition source, ventilation conditions, environmental conditions, etc.), it would be obvious how to test it for fire hazard. One must evaluate a series of candidate materials under a particular set of full-scale fire circumstances and note which materials are satisfactory or unsatisfactory. One would then evaluate smaller samples of the same series of materials by a proposed test method, which would become validated for future use if the results of the two procedures correlated.

Frequently, this idealized approach is not directly applicable because a material of interest may be exposed to fire risk in a variety of ways rather than in a simple well-specified set of circumstances. Thus, an incredibly large number of experimental burns would be required to explore fully all the permutations and combinations of the variables. A second reason is that in some cases even a few realistic experimental burns would not be feasible; e.g., if each experiment required destruction of a high-rise building or a Boeing 747, or, if an experiment required exposure of human beings to lethal fumes.

Many fire test methods, lacking full-scale validation, are highly unreliable indicators of hazard. For many cases, no appropriate test methods exist. Tests are particularly unreliable with regard to fire spread and growth, as well as noxious gas and smoke production.

If the scientific studies of combustion were sufficiently advanced, one might hope to turn to scientists for the development of theoretical links between physical and chemical phenomena involved in test methods and the corresponding phenomena involved in real fires. Such theoretical knowledge, even if incomplete, would appear to be useful in generalizing from limited fire experience; it would be a powerful supplement to empiricism. However, combustion scientists have made rather limited progress in untangling the complexities of fire behavior and are unable at this time to analyze most fire situations in fundamental terms. Research, involving highly idealized materials and geometries, is progressing on a limited scale. In other research efforts, some realistic fires are highly instrumented to obtain useful information, but much additional effort is required before we can expect major contributions by combustion scientists to more relevant test methods. The importance of fundamental research in this area is discussed further in Volume 4 of this series, "Fire Dynamics and Scenarios," Chapter 2 and 3.

6.2 The Scope of Fire Dynamics

What is fire dynamics? In broad terms, it involves quantitative descriptions of fire phenomena on a scientific basis. An understanding of these phenomena may be obtained on any one of four levels, as follows:

On the most basic level, we consider fire to be governed by a combination of effects tracing back to certain established fields of scientific study such as: the chemical thermodynamics and stoichiometry of combustion; the chemical kinetics of pyrolysis and combustion reactions; the transfer of combustion energy by conduction, convection, and radiation; the motion of combustion gases as affected by buoyancy, thermal expansion or mechanical force, and as modified by constraining walls, viscous effects, inertia, turbulence, the properties of hot gases, and other pertinent parameters.

These factors are of interest for application purposes in addition to fire dynamics, but are considered to be proper subjects for academic rather than industrial research. Knowledge deriving from a study of these factors constitutes the foundation of fire dynamic studies.

On the second level, we may break fire down into a series of phases or stages, and consider fire dynamics to be an analysis of any of these phases:

smoldering
spontaneous ignition
piloted ignition
horizontal or downward flame spread over solids
    (thermally thin or thermally thick)
upward flame spread over solids
flame spread over a liquid below its flash point
flame spread over a liquid above its flash point
burning rate of a liquid pool
burning rate of a solid slab
    (charring or non-charring)
    (vertical or horizontal, facing up or down)
    (laminar or turbulent flame)
formation of toxic species in a diffusion flame
formation of aerosols in a diffusion flame
radiation emitted by a diffusion flame
extinguishment by heat loss
    (radiation, convection, heat-absorbing substance)
extinguishment by reduction of oxygen
extinguishment by chemical inhibitors

Research in these phases is conducted largely but not entirely in academic institutions; results would be of general interest to persons beyond the fire community concerned with combustion, such as furnace designers and air pollution control administrators.

On the third level, we consider fire dynamics to be concerned with complex processes generally involving interactions of two or more phases given in the foregoing list. Some examples of such fire dynamics are as follows:

- burning rate of an object as influenced by radiative feedback from the environment
- burning rate of an object in a non-combustible compartment as influenced by ventilation of the compartment
- generation of incomplete combustion products as influenced by either of the above burning rate conditions
- mutual interactions of two adjacent burning objects
- spontaneous ignition of pyrolysis gases from a hot object as influenced by turbulent free convection and mixing
- properties of smoke from a fire in a compartment as influenced by the mixing, cooling, and "aging" or agglomeration, which occurs in the interval between smoke generation and its arrival at a detector station
- adsorption of toxic gases from fires by aerosols from the same fires, as cooling occurs
- analysis of radiant emission, transmission, and absorption in a compartment at a pre-flashover stage of fire
- the effect of physical scale on fire turbulence, on fire radiation, and ultimately on fire behavior
- delineation of relative effects of chemical kinetics and physical factors on a fire near extinguishment conditions
- determination of the source of undesirable products of incomplete combustion in a fire; i.e., surviving initial pyrolysis products vs. products formed in gas-phase reactions in or near a flame
- effects of long-term exposure of materials to various ambient conditions on subsequent fire behavior
- identification of the mechanisms involved in the interaction of water spray with a fire, e.g., cooling of pyrolyzing solids, cooling and/or diluting of flame with steam, prewetting of adjacent fuel, absorbing radiation, entering into the flame chemistry via $C + H_2 + O$ or $CO + H_2O$, entraining air with the water spray, as well as exploding superheated droplets in molten polymer and spattering fuel that feeds the combustion process.

Obviously, successful fire dynamics studies along these lines would strengthen the engineering judgment needed to validate test methods for the realistic determination of fire hazards. Academic researchers tend to avoid this third level of problems, because many of them feel that so many major unknowns in the second and first level problems exist that third level problems are still too difficult to treat properly.

So far, except for mentioning toxicity, we have been discussing fire as if it occurs in an uninhabited world. Consequently, the fourth level of fire dynamics involves interactions of fire phenomena with human response. Problems such as the following exist:

– sensory detectability of fire (including smell and sound)
– vision as affected by smoke and/or lachrymatory gases
– panic or confused thinking as induced by fire phenomena
– burn damage of skin by garments, particularly as influenced by the rate of flame spread, melting-dripping, etc.
– toxicity including combined (synergistic) or sequential effects of various toxic species
– toxicity, including prefire condition of victim (blood-alcohol content, heart or circulatory disease, etc.)
– human ability to control fire at various stages of development as governed by training, equipment available, panic, etc.
– actions of humans which lead to ignition

It is recognized that the study of human behavior is outside the scope of this study, but it seems reasonable to call attention to this important aspect of the problem to complete the catalog of what we need to know to obtain a complete understanding of fire dynamics.

While comprehensive, the foregoing four levels of approach to fire dynamics exclude phenomena related to the post-flashover development of a fire. They do not consider the spread of a fire from one part of a building to another; the interaction of building components (ventilation system, windows, elevators, etc.) with a fire; the spread of fire from one structure to an adjacent structure; and technology concerned with detection, communication, escape procedures, etc. Although a strictly scientific study of some elements of these factors is possible, engineering or systems analysis, involving largely empirical and state-of-the-art knowledge, is primarily required. Accordingly, the Committee feels that study of such complex processes should not be called fire dynamics, even though such studies should utilize fire dynamics when possible.

6.3 Current State of the Art of Critical Fire Dynamics Elements

6.3.1 Introduction

In recent years, most current research on fire dynamics in the U.S. has been funded to various grantees by the RANN program of the National Science Foundation. However the Federal Fire Prevention and Control Act of 1974 assigned responsibility for "basic and applied research for the purpose of arriving at an understanding of the fundamental processes underlying all aspects of fire" to the Center for Fire Research of the National Bureau of Standards. The Center conducts an internal program and is expected to provide much of the future support of external fire dynamics research. Factory Mutual Research Corporation (FMRC) supports its own internal program of fire dynamics research. Several other government agencies support some fire research which includes, to a minor degree, fire dynamics studies. Table 6-1, based on estimates, is a representation of the total

Table 6-1. Fire Dynamics (FD) Research as Budget in U.S. Programs (Millions of Dollars)

FISCAL YEAR	NSF-RANN FIRE PROGRAM	NFPCA RES. PROGRAM	NSF + NFPCA			NBS CFR			FMRC INTERNAL PROGRAM			MISC. FD*	TOTAL FIRE DYN. RES.
			TOTAL	%FD*	TOTAL FD	TOTAL	%FD*	TOTAL FD	TOTAL	%FD	TOTAL FD		
73	2.00	0	2.00	60%	1.20	1.9	10%	0.19	1.0	50%	0.50	0.50	2.39
74	1.70	0	1.70	60%	1.02	3.1	10%	0.31	1.0	50%	0.50	0.50	2.33
75	0.65	1.12	1.77	60%	1.06	3.1	10%	0.31	1.0	50%	0.50	0.50	2.42
76	1.16	1.20	2.36	60%	1.42	3.1	10%	0.31	1.0	50%	0.50	0.50	2.73
77	0	1.50	1.50	20%	0.30	4.5	15%	0.72	1.0	50%	0.50	0.50	2.02

*Estimated

funding for fire dynamics research (research on the toxicology of fire gases is not included). It shows that the total available fire dynamics funding increased from \$2.39 million in FY 1973 to \$2.73 million in FY 1976, but, as of February 1976, is scheduled to decrease to \$2.02 million in FY 1977.

Table 6.2 lists the major U.S. research organizations performing fire dynamics research. In about half the cases, however, the activity consists of a single professor working part-time with one or two graduate students.

Table 6-2. Some U.S. Research Organizations Performing Fire Dynamics and Related Research in 1975

(in roughly approximated decreased order of activity level)

1. Center for Fire Research, National Bureau of Standards, Gaithersburg, Md., Dr. J. Lyons, Dr. R. S. Levine, Dr. J. Rockett, Dr. C. Huggett, Dr. J. Quintiere, Dr. T. Kashiwagi, Dr. H. Baum, Dr. R. J. McCarter.
2. Factory Mutual Research Corporation, Norwood, Mass., Dr. R. Friedman, Dr. J. de Ris, Dr. G. Heskestad, Dr. G. H. Markstein, Dr. R. L. Alpert, Dr. P. A. Croce, Dr. A. Modak, Dr. F. Tamanini, Dr. A. Tewarson.
3. University of California, Berkeley, Cal., Professor R. B. Williamson, Professor C. L. Tien, Professor P. J. Pagni, Professor R. Sawyer.
4. University of Utah, Salt Lake City, Utah, Professor J. D. Seader, Professor N. W. Ryan, Professor I. Einhorn.
5. Applied Physics Laboratory, Johns Hopkins University, Silver Spring, Md., Dr. W. G. Berl, Dr. R. M. Fristrom.
6. Harvard University, Cambridge, Mass., Professor H. W. Emmons.
7. Massachusetts Institute of Technology, Cambridge, Mass., Professor T. Y. Toong, Professor J. A. Fay, Professor G. C. Williams.
8. Princeton University, Princeton, N. J., Professor I. Glassman, Professor M. Summerfield, Professor W. A. Sirignano, Dr. F. L. Dryer.
9. Georgia Institute of Technology, Atlanta, Ga., Professor B. T. Zinn, Professor P. Durbetaki.
10. University of Notre Dame, South Bend, In., Professor K. Yang, Professor J. R. Lloyd, Professor M. L. Doria.
11. University of California, San Diego, Cal., Professor F. A. Williams.
12. Pittsburgh Mining and Safety Research Center, U.S. Bureau of Mines, Pittsburgh, Pa., Dr. D. Burgess.
13. Brown University, Providence, R. I., Professor M. Sibulkin.
14. Cornell University, Ithaca, N. Y., Professor K. Torrance.
15. Stevens Institute of Technology, Hoboken, N. J., Professor R. S. Magee
16. University of Montana, Missoula, Montana, Professor F. Shafizadeh.
17. Northwestern University, Evanston, Ill., Professor M. C. Yuen.
18. California Institute of Technology, Pasadena, Cal., Professor E. Zukoski.
19. State University of New York, Stony Brook, N. Y., Professor R. Lee, Professor A. L. Berlad.
20. University of Maine, Orono, Maine, Professor A. Campbell.
21. University of Washington, Seattle, Wash., Professor R. Corlett.
22. Pennsylvania State University, State College of Pennsylvania, Professor G. Faeth.

A rather complete (although not quite current) guide to the published literature relevant to fire dynamics is *Fire Research Abstracts and Reviews,* published several times each year since 1958 by the National Academy of Sciences. Although discontinued in 1977, the *Fire Research Abstracts and Reviews* are available through Vol. 18, 1976. Another source is *References to Scientific Literature on Fire,* published annually for many years by the Joint Fire Research Organization, Borehamwood, Herts., England. The following journals contain occasional relevant papers: Combustion and Flame, Combustion Science and Technology, Journal of Fire and Flammability, and Fire Technology. See also the biennial International Symposia on Combustion (The Combustion Institute, Pittsburgh). Two new journals, Fire and Materials and Fire Research, commenced publication in 1976. Finally, some relevant books (Lewis and von Elbe, 1961; Williams, 1965; NAS-NRC, 1961; N.B.S., 1972) are referenced.

To illustrate the current state of the art, several samples of fire dynamics studies and their importance to test method development are as follows:

6.3.2 Burning Rate of Plastic Slab

Once fully ignited, a thick plastic slab burns at a rate which is believed to be essentially independent of all chemical kinetic parameters, except those that might affect the luminosity of the flame. The theory relating this rate to the governing parameters is well advanced (Friedman, 1971).

For the case of a small or nonluminous flame, radiation is negligible, and the burning rate is controlled by an energy balance at the surface, in which heat is convected from the flame gases to the surface and absorbed by conduction into the interior, endothermic heat-up, depolymerization, and gasification of the plastic (some of the heat is lost by re-radiation from the surface). While depolymerization reactions obey the laws of chemical kinetics (i.e., proceed with finite activation energies), the kinetics has very little influence on the rate of gasification because the surface assumes a sufficiently high temperature to permit the reactions to occur just fast enough to satisfy the energy balance.

Furthermore, the gaseous flame reactions are describable by the Burke-Schumann diffusion flame model, or refinements thereof, in which the reaction rates are assumed to be infinite, and heat transfer as well as diffusion rates control the process.

For the case of laminar flow and negligible radiation, a complete numerical theory has been formulated. It agrees with experiment (Kim, deRis, and Kroesser, 1971). The important parameters are the Spalding B-number (Friedman, 1971), which is essentially the ratio of the heat of combustion to the heat of gasification times the stoichiometric fuel-air ratio, and the Grashof number, which is a ratio involving buoyancy force acting on an element of hot gas and viscous force. Weaker parameters are the surface temperture, the gas transport properties, and the stoichiometric ratio (in addition to its effect on the B-number).

When the specimen has dimensions of about 10 cm or more, the flame will be turbulent instead of laminar and large enough so that radiation from the flame to the surface becomes significant. Neither the theory of turbulence nor the theory of flame radiation is sufficiently advanced to permit quantitative calculation of burning rate in this case from first principles, as contrasted with the preceding laminar case. However, experimental studies of such turbulent flames have indicated the relative importance of radiation.

For example, a horizontal slab of polystyrene, 18 cm square, burning as a pool, is consumed nearly twice as fast (mass per unit area) as an identical slab of polyoxymethylene (Magee and Reitz, 1975). In contrast, when a small specimen of polystyrene rod is burned in an apparatus with an opposed air jet (Holve and Sawyer, 1975), it is consumed at less than half as fast as polyoxymethylene. When burning as a large turbulent pool, polystyrene has a flame which is highly luminous, whereas a polyoxymethylene flame is blue. In small opposed-jet burners, both flames are blue.

Additionally, when a large polymethyl methacrylate slab is burned vertically (on one side) instead of horizontally, the burning rate is found to be lower. An explanation for this phenomenon is that a point on the surface of the vertical slab doesn't "see" as great a thickness of flame as the corresponding point on the horizontal slab, because the flame shape is different and, accordingly, less radiation is received. It has been possible to make direct measurements of the radiant and convective heat arriving at the surface of a vertical polymethyl methacrylate slab; at a point 50 cm from the bottom, the incident radiant flux is 2.6 times as large as the convective flux from the flame (Orloff, de Ris, and Markstein, 1975). Thus, even larger radiant feedback would be expected from the horizontal burning configuration.

Furthermore, the rate of increase in burning rate for burning plastics, with incident radiant energy supplied by electrical radiant panels, has been measured (Magee and Reitz, 1975). For example, a radiant flux of only 1.3 watts/cm^2 increased the steady burning rate of a vertical polymethyl methacrylate slab by a factor of 2.5. A ceiling of unit emissivity would emit such a radiant flux if it reached a temperature of 413°C. Large fires emit radiant fluxes at least five times as great as this value. Clearly, this effect can be of dominant importance.

Burning rates of thick slabs of wood or charring plastics are much lower than those of melting plastics, for two main reasons: 1) the char layer acts as insulation between the flame and the virgin fuel; 2) since the surface temperature is about 900 K instead of about 650 K, the re-radiation of heat from the surface, which varies as the fourth power of surface temperature, is much higher.

These recently acquired insights into the role of radiation in burning give promise for a better understanding of these phenomena. Any flammability test method must take into account the importance of radiative transfer in fires.

6.3.3 Burning in an Enclosure

A wood crib in the open burns at a reproducible rate, which may be correlated with the geometrical parameters of the crib (Block, 1971). When such a crib is burned in a non-combustible enclosure with minimal ventilation, the burning rate is greatly reduced, and the proportion of carbon monoxide in the combustion products is greatly increased (Tewarson, 1972). However, if the test were repeated in an enclosure that is very well ventilated, the burning rate would be considerably higher (up to 70% greater) than the burning rate for the same crib in the open (Croce, 1976). A possible explanation for this effect is that the ceiling above the crib gets hot and radiates heat back to the crib. It is also possible that the hot smoky gases layered under the ceiling also radiate significant heat to the crib. While the relative radiative effects of the ceiling and the hot gases under the ceiling have not yet been differentiated, it is clear that the radiant feedback is important.

In a more dramatic experiment, a horizontal slab of polymethyl methacrylate, 30 inches square, was burned under a ceiling 4 feet high, with adequate ventiliation. The slab burned for a short initial period at a rate corresponding to burning in the open, but then rapidly accelerated to a rate about four times as great, as radiant feedback built up (Croce, 1976).

Since the burning of materials in compartment is common to most fires, further development of the laws governing such burning is essential. Any flammability test methods designed to measure the hazard contribution by individual items in a compartment, e.g., upholstered chair, must account for the effect of the compartment on burning behavior.

6.4 Applications of Fire Dynamics to Test Method Development

In general, our knowledge of fire dynamics is still primitive; it is difficult to point to many examples of its use in test method development. Most tests were developed when such knowledge was elementary, or at times when enormous pressure existed to develop test methods in a matter of months, permitting no time for a realistic fire dynamics approach. Even now, tests are often introduced in response to a specific disaster rather than as part of a long-range development plan.

However, in a recent case, fire dynamics considerations were used in test development for fire spread in a corridor with flammable floor covering. In this case, corridor fire experiments were performed, and the radiation from the ceiling as well as from the burning gases just below the ceiling was identified as critical in promoting a "flameover condition" (propagation rates of the order of a foot per second) over the floor covering. Thus, a test is being developed involving radiant heat impinging on a floor-covering sample, with piloted ignition at one end (see Figure 6.1). In that test, the flame starts in a region of high radiation flux and propagates to regions of progressively lower radiant flux; then, at some point, the flame ceases to advance. The acceptance criterion would be the radiant flux level

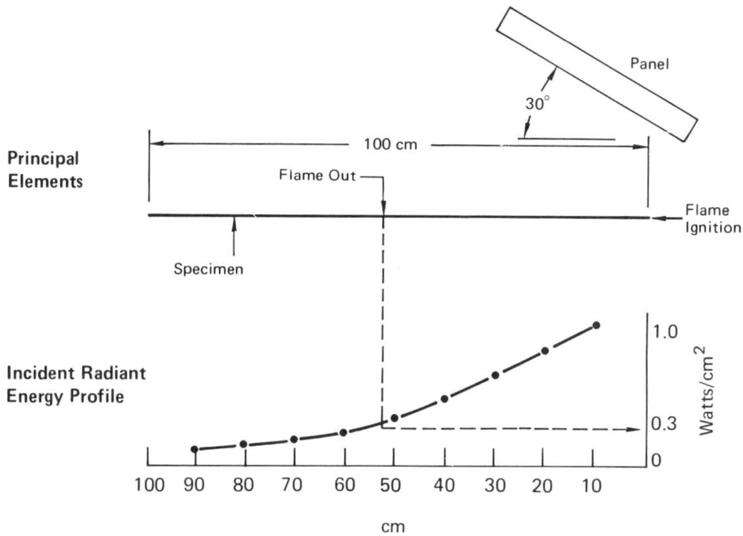

Figure 6.1. Test of radiant heat impinging on floor-covering material.

below which propagation must not occur. This critical flux criterion, approximately 0.25 to 0.5 watts/sq cm, would be determined by the occupancy under consideration. These fire dynamics studies (Hartzell, 1974; Denyes and Quintiere, 1973; Quintiere, 1975a; Kashiwagi, 1975; Quintiere, 1975b; Benjamin and Adams, 1976) established the importance of this variable and provided guidance as to a proper value for the critical flux.

Another approach to test method development is the use of fire modeling. Some historical background on this subject was presented in a 1959 symposium (Berl, 1961). Also see references (Kashiwagi, 1975; de Ris, Kanury and Yuen, 1973; de Ris, 1973; Emmons, 1973). In the fire modeling approach, a physical model of a fire situation is reduced in scale while maintaining geometric similarity and preserving the important chemical and thermodynamic properties of the materials involved. Were such an approach successful, the benefits to fire testing would be very great. Unfortunately, progress in its development has been very slow.

The difficulties of modeling a fire by reducing scale arise in several ways: 1) very small fires are laminar while larger fires are turbulent; 2) as far as fluid mechanics is concerned, the ratio of buoyancy forces to viscous forces in the convective flow of fire gases is size-dependent; 3) the radiant emission and self-absorption of the flame are size-dependent; 4) the gas-phase time scale in the fire is shorter for small than for large fires, with possible effect on incomplete combustion products. These formidable difficulties have kept people from having much confidence in fire model test results.

However, it is anticipated that valid modeling procedures may be developed for at least some aspects of fire behavior as the understanding of fire dynamics relevant to situations of interest improves. Then, compensation for any errors or distortions introduced by the modeling might be achieved by varying other parameters such as ambient temperature, pressure, oxygen concentration, or ambient radiation. For example, it has been suggested that the gravitational force may be varied by use of a centrifuge.

Some progress has been made by a pressure-modeling technique (Berl, 1961), which is based on the principle that the Grashof number is invariant if the product of pressure squared and size cubed is held constant. The Grashof number is a dimensionless ratio of buoyancy forces to viscous forces in the hot gases around a fire. Convective heat transfer from the fire plume to the surroundings, expressed as a dimensionless quantity, is governed by the Grashof number. Pressure-modeling has been shown to be valid for diffusion-controlled burning as long as radiation either varies with pressure in certain prescribed ways or is negligible. Pressure-modeling has worked over a tenfold specimen size range for turbulent burning of polymethyl methacrylate and wood objects in several shapes. It is also possible to model the temperature and velocity distributions near a fire in a building, by using a reduced-scale physical model at atmospheric pressure (if the fire intensity is controllable) and reducing the fire intensity so as to maintain a constant Froude number (the ratio of the square of the maximum fire plume velocity to the ceiling height). (de Ris, Kanury, Yuen, 1973). This modeling procedure appears to be valid except in the viscous boundary layer. Future progress in the modeling of fire situations will be difficult to achieve without a better understanding of the underlying controlling mechanisms of fire.

6.5 Conclusions and Recommendations

Conclusion: The complex dynamic interaction of fundamental processes in fire dynamics, coupled with a relatively small funding effort, has led to rather slow progress in quantifying fire behavior. *Recommendation:* An increased program of sufficient magnitude and stability, funded by the government, in the specific area of fire dynamics, must be established. This program should be conducted primarily in academic, non-profit, and governmental research organizations, but the projects should be closely monitored by advisory boards including broad representation by manufacturers, users, and members of standards-setting organizations.

Conclusion: Fire dynamics research funding for FY 1977 is estimated to *decrease* 26% below the FY 1976 level, because of federal government budgetary actions. An *increase* of perhaps 8% is needed even to maintain the past level of effort, because of inflation. *Recommendation:* At least an additional 50 scientists and engineers (approximately equal to the current effort) should be supported over a 10-year period in the specific area of fire dynamics research. This increase would represent an additional cost of approximately $2.5 million per year; approximately

0.02 percent of the annual cost to society of fires in the United States. (President's Commission on Fire Protection and Control estimate for 1972 was some $11.4 billion). In view of the shortage of qualified engineers and scientists in the appropriate disciplines, the program should be brought to full strength within the first four years.

Conclusion: Our current level of understanding of fire dynamics is not sufficiently developed to permit thorough scientific analysis of most fire test methods. The behavior of a given material in a fire is dependent not only on the properties of the fuel, but also on the fire environment to which it is exposed. Consequently, if test methods are to be meaningful, they must simulate the critical fire dynamic conditions. An understanding of fire dynamics is essential if the critical conditions are to be identified. Consequently, fire dynamics can be extremely useful in test method selection and development. Valid modeling procedures, based on evolving fire dynamic principles, offer promise for reduction in the dependence on, and the cost of, large, scale fire tests. *Recommendation:* Fire dynamics expertise should be employed when developing new test methods and for validation and improvement of existing test methods.

CHAPTER 7

SPECIFICATIONS, STANDARDS, CODES

7.1 Introduction and Purpose

The technical status of methods for evaluating flammability behavior of polymeric materials has been reviewed and discussed in the preceding chapters. Clearly, test methods play an essential role in fire research, and in the development of improved materials. However, test methods do not in themselves provide the desired improvements in fire safety unless they are used in conjunction with enforced specifications, regulations or standards. The purpose of this chapter is to indicate the most important sources for the issuance and enforcement of regulatory documents. These generally include reference to or description of specific test methods, and of criteria according to which a given material can be claimed to comply. In the United States, there are many sources of regulations, and many routes leading to the promulgation of standards. Test methods specified in the regulations may have been developed *de novo* for the material or product covered, or they have been based on existing techniques with appropriate modifications as needed. Pass/fail criteria are established in the framework of existing technology in the course of the regulatory process.

7.2 Definitions and Word Usage

The terms employed in this chapter have been discussed in Chapter 3 (Overview of test methods, standards, specifications and codes) which should be referred to.

7.3 Origin of Documents

7.3.1 Private Institutions

Specifications and standards including reference to specific test methods and procedures are frequently developed by technical groups in private institutions in order to define properties (including flammability behavior) of materials of interest to the institution. Examples are: *Industrial organizations* which either produce and market, or purchase and use polymeric materials; *Trade associations* which reflect the needs and interests of producer or user companies; *Professional and technical societies* which may develop test methods through committees, and provide a technical base for the evaluation of materials; *Insurance companies*, which are concerned with prevention of fires with reference to their clients. Documents which originate in private institutions may be simply recorded in the technical literature, or they may become a part of handbooks issued annually by technical societies, and widely used as reference in industry (e.g., American Society of Testing and Materials – ASTM; American Association of Textile Chemists and Colorists – AATCC; Society of the Plastic Industry – SPI).

Mention should also be made of the American National Standards Institute (ANSI), which is a federation of technical, trade, industrial, professional and labor organizations working with government agencies and consumer groups to coordinate standards development in the private sector.

7.3.2 Voluntary Standards

In a recent publication (National Bureau of Standards, 1975), 506 technical, nontechnical, trade and professional organizations have been identified as generating voluntary standards; 116 organizations are listed in the area of safety, and about 25 organizations (National Building Code, 1976) relate to fire safety. The existing voluntary standardization system in the United States is discussed in detail in a 1977 report of the National Materials Advisory Board.

Organizations involved are classified as:

A. Voluntary Standards Writing and Promulgating Bodies (e.g., ASTM, ANSI);
B. Professional Societies;
C. Trade Associations;
D. "Listing" Bodies; and
E. Scientific Bodies

Standardization procedures of the organizations of major importance (e.g., ASTM) are discussed in (National Bureau of Standards, 1975). Certification of compliance by producers is essential to the acceptance and observance of certain types of voluntary industrial standards, and provides guidance and protection to the buyer. The extent of testing and inspection required to assure compliance with the standard are important factors in determining the value of certification.

The role of voluntary standards in the mandatory standard process has been recently addressed by the Consumer Product Safety Commission (U.S. Consumer Product Safety Commission, 1977); it has been stated that "a proper combination of voluntary and mandatory standards can bring a greater 'pay-off' in increased product safety than either type of standard alone."

7.3.3 Codes

Codes are compilations of mandatory standards which are used for regulation of many aspects of life. Codes relevant to the subject of this chapter include: *Fire prevention codes* and *Building codes* (Fire Prevention Handbook, 1976).

The National Fire Protection Association (NFPA) is a principal source of consensus standards and codes for fire prevention. (Many of these have been incorporated into law by government at various levels).

With the participation of as many as 2500 experts from many fields, the NFPA develops fire safety standards and codes which are published for voluntary adoption by any organization (private or public) or jurisdiction with power to enforce. These standards are published both individually and in a set of several volumes called *National Fire Codes.* They are periodically revised, based on the findings and activities of the many technical committees of NFPA.

In addition, the NFPA provides technical advisory services to anyone having a legitimate concern for fire safety. This service is furnished through technical meetings and publication of technical books, pamphlets, and periodicals. The NFPA *Fire Protection Handbook* is recognized as the most extensive reference on fire protection, and includes digests of federal and state laws affecting fire safety.

Four *model building codes* have been promulgated in the United States: the National Building Code, the Basic Code, the Standard Building Code, and the Uniform Building Code.

While the National Building Code, published by the American Insurance Association (AIA) (formerly the National Board of Fire Underwriters), is prepared by the AIA Technical Staff, the other three model codes are published by organizations representing building officials, with their voting memberships having the final determination on the contents of the codes. These organizations are (1) Building Officials and Code Administrators International Inc. (Basic Building Code), (2) Southern Building Code Congress International Inc. (Standard Building Code), (3) The International Conference of Building Officials (Uniform Building Code). Each of these three organizations representing building officials also publishes one or more other building-related codes such as mechanical, plumbing, fire prevention and the like, which are coordinated with the building code, for the purpose of eliminating discrepancies in the codes of a particular organization.

7.3.4 State and Local Regulations

To date, only about half the states have formal activity in the promulgation of standards for fire safety. Standards activities in many states, cities and other governmental divisions are principally associated with purchasing; specifications and standards are drafted in conjunction with procurement documents; engineers may or may not have been involved in identifying the test methods or criteria to be used in the documents. There are, however, many exceptions, e.g., the State of California (State of California, 1975).

7.3.5 Federal Documents

Federal documents of interest fall into three classes: (1) *procurement documents,* (2) *guideline documents,* (3) *regulatory documents.* The first and second groups set the basis for bidding by suppliers to the Federal Government. Procurement documents are prepared primarily by the General Services Administration and by the Armed Services, although, theoretically, any part of the government may prepare specifications for goods and services that it desires to acquire. It is usual, though not required, for the bulk of a government specification to be drawn from a previously accepted document originating either in a government agency or in a private source.

The third group of federal documents includes all mandatory standards issued and enforced by the Federal Government.

A study has been made for the Department of Defense on Materials and Process Specifications and Standards. Despite its defense emphasis, it gives a good overview of current specification and standard activities in the Federal Government. The reader is referred to this study for details on this subject (National Materials Advisory Board, 1977).

The promulgation of a federal mandatory standard is not a simple process. It starts with a series of hearings, sometimes conducted by committees of Congress or by the specific government agency or department which has jurisdiction. This is followed by publication in the Federal Register, of a statement indicating intent to issue a standard, or of a Notice of Finding that a standard may be needed. This is in turn followed by publication of a draft of the document, solicitation of comments, proposed changes, further hearings, and finally publication of the Standard in the Federal Register.

Government departments and agencies that are concerned with some aspects of regulations on flammability and fire safety of materials are partially listed in Table 1.

7.4. Enforcement of Standards and Regulations in the U.S.

The standards system of the United States is highly fragmented, unlike that of other developed nations. For example, mandatory building fire codes are enacted and enforced at the local level (town, country, city). The Flammable Fabrics Act is enforced by the U.S. Confumser Product Safety Commission; fire hazards in mines are the province of the Mine Health and Safety Administration (Department of Labor); fire hazards in nuclear power plants are the concern of the Nuclear Regulatory Commission; fire hazards on commercial vessels are the concern of the U.S. Coast Guard (U.S. Department of Transportation); fire safety in aircraft is a joint care of the Federal Aviation Authority (U.S. Department of Transportation) and the National Aeronautics and Space Administration; the Veterans Administration is concerned with fire safety in its hospitals; the General Services Administration is concerned with fire safety in federal buildings. The list could be much longer, since the United States does not have a central agency dealing with fire safety, nor does it have a single central enforcement agency. The coordination of fire safety activities in federal agencies, however, is assigned by the National Fire Protection and Control Act of the National Fire Prevention and Control Administration of the Department of Commerce.

At the local level, fire safety regulations are enforced through the power of building permits, licensing authority, and inspections which are required for occupancy of buildings. Building and operation of vehicles may be permitted or not

Table 1. Government Departments and Agencies Concerned with Regulations on Flammability and Fire Safety of Materials

Department	Agency/Administration
Department of Commerce	Maritime Administration Office of Product Standards National Bureau of Standards National Fire Prevention and Control Administration (NFPCA)
Department of Defense	Army Navy Air Force Defense Supply Agency
Department of Health, Education, and Welfare	National Institute for Occupational Safety and Health
Department of Housing and Urban Development	– –
Department of the Interior	Bureau of Mines
Department of Labor	Occupational Safety and Health Administration Mine Safety Enforcement Administration
Department of Transportation	United States Coast Guard Federal Railroad Administration Federal Aviation Administration National Highway Traffic Safety Administration
Independent Agencies	Consumer Product Safety Commission (CPSC) General Services Administration National Aeronautics and Space Administration Veterans Administration

permitted according to the degree of conformity with established fire safety standards: This applies to ships, aircraft, and land vehicles. Police power in this area can be also exercised by the state (notably California) but is more generally exercised by the Federal Government.

Economic incentives can also have influence in the drive towards fire safety. Insurance premiums, for example, can be inversely proportional to estimated fire safety. On the other hand, a recent study (Erling and Reiser, 1976) has shown that "In America, fire department costs are typically funded from property taxes. This has resulted in a negative tax incentive situation which discourages the use of fire safe construction and installation of private fire protection equipment in buildings. Large buildings without on-site fire protection installations require public fire departments to focus an inordinate amount of resource to deal with the high fire flow requirement. This usually results in the small building owner subsidizing the fire department cost to structures other than his own." The study proposes a way to reward property owners for providing fire protection to their property.

7.5 Conclusions and Recommendations

Conclusion: The major importance of test methods is that they become a part of the specifications, standards, and codes which define the performance of materials. *Recommendation:* Test methods incorporated in regulations must be carefully investigated to establish their validity and limitations.

Conclusion: Only the United States among developed nations has a widely fragmented system of fire safety standards and codes. This contributes to losses of life and property which might be avoided. *Recommendation:* Attack the problem of fragmentation of fire codes by judiciously evaluating the modes used in other developed countries and adopting their practices as seems desirable in the United States.

Conclusion: Codes and standards are only of value to the extent that they are enforced. *Recommendation:* Enforcement of fire codes and mandatory standards should be uniform and rigorous.

Conclusion: Recognizing that all contributors to the fire protection system should be stimulated to maximum activity in the pursuit of fire safety, incentives should be sought to further these activities. *Recommendation:* Search out economic incentives such as favorable insurance rates, favorable tax treatment, or other societal incentives that would favor improved fire safety.

REFERENCES

Adams, J. S., Williams, D. W., and Tragellas-Williams, J.; "Air Velocity, Temperature, and Radiant-Heat Measurements Within and Around a Large Free-Burning Fire;" Fourteenth Symposium (International) on Combustion, P. 1045–1052, The Combustion Institute, 1973.

Alpert, R. L.; "Pressure Modeling of Transient Crib Fires;" Factory Mutual Research Corporation Serial No. 22360-2; December 1975.

Alpert, R. L., Modak, A. T., and Newman, J. S.; :"The Third Full-Scale Bedroom Fire Test of the Home Fire Project, Volume 1 – Test Description and Experimental Data;" Factory Mutual Research (July 30, 1975).

Amaro, A. J., Kanury, A. M., Lipska, A. E., and Martin, S. B.; Thermal Indices from Heat-Release-Rate Calorimetry, Paper No. 37, Presented at the Western States Section of the Combustion Institute Fall Meeting, 1974.

American Society for Testing and Materials, "Guide for Room Fire Experiment," ASTM E 603-77, 1977.

Annual Book of ASTM Standards; American Society of Testing Materials, Philadelphia, Pa., Vol. 18 (Note: All ASTM referenced tests can be found in either Vol. 18 or Vol. 35), 1978.

"Approval Standard for Air and Fume Handling Ducts" FMRC Std. 4922, Factory Mutual Research, Norwood, Mass. (March 1973).

Babrauskas, V., and Williamson, R. B.; "Post-Flashover Compartment Fires;" University of California, Berkeley, Report No. UCB FRG 75-1, December, 1975.

Benjamin, I. A., and Adams, C. H.; "The Flooring Radiant Panel Test and Proposed Criteria," Fire Journal *70,* p. 63, (1976).

Benjamin, I. A., and Adams, C. H.; "The Flooring Radiant Panel Test and Proposed Criteria," Fire Journal, *70* (2), p. 63–70, 1976.

Benjamin, I. A.; "Problems in the Correlation of Small and Large Scale Tests;" International Symposium on Fire Safety of Combustible Materials; University of Edinburgh, p. 141–147, October, 1975.

Berl, W. G. (Editor), "The Use of Models in Fire Research," National Academy of Sciences-NRC., Publication 786, Washington, D.C., (1961).

Block, J. A., "Nonpropagating Free-Burning Fires," Thirteenth Symposium (International) on Combustion, The Combustion Institute, pp. 971–978, (1971).

Braun, E.; Cobble, V. B., Helzer, S., Krasny, J. F., Peacock, R. D., and Stratton, A. K., NBS Report No. NBSIR-76-1072, June 1976.

Braun, E.; "Report of Fire Test on an AM General Metro Bus;" NBSIR 75-718; National Bureau of Standards, June, 1975.

Brenden, J. J.; An Apparatus Developed to Measure Rate of Heat Release From Building Materials, Research Paper FPL 217, Forest Products Laboratory, U.S. Department of Agriculture, Madison, Wisconsin, 1975.

Bryson, J. O., Gross, D.; Techniques for the Survey and Evaluation of Live Floor Loads and Fire Loads in Modern Office Buildings (Bldg. Sci. Ser. 16). National Bureau of Standards, Washington (1967).

Buchbinder, B., and Buchbinder, L. E.; "The Fire Hazard Analysis Process at the National Bureau of Standards," International Symposium on Fire Safety of Combustible Materials; University of Edinburgh, p. 156–161, October, 1975.

Butler, D. P., Martin, S. B., and Wiersma, S. J.; "Measurement of Dynamics of Structural Fires;" Fourteenth Symposium (International) on Combustion, p. 1053–1061, 1973.

Castino, G. T., Beyreis, J. R., and Metes, W. S.; "Flammability Studies of Cellular Plastics and

Other Building Materials Used for Interior Finishes," Underwriters' Laboratories, Inc., June 13, 1975.

Christian, W. J., and Waterman, T. B.; "Fire Behavior of Interior Finish Materials," Fire Technology, Aug. 1970.

Chamberlain, D.; Rate of Heat Release – A Tool for the Evaluation of the Fire Performance of Materials, Presented at the Joint Meeting of the Central States and Western States Section of the Combustion Institute on Flammability and Burning Characteristics of Materials and Fuels, San Antonio, Texas, April 21–22, 1975.

Chrysler Corp., Lab. Proc. LP-463KC-13-01 (November 14, 1968).

Clarke, F. B., and Ottoson, J.; "Fire Death Scenarios and Firesafety Planning," Fire Journal, *70,* 20, 1976.

Corporation Report FMRC Serial No. 21011.6, October, 1975.

Corson, R. C., and Lucas, W. R.; Life Hazard of Interior Finishes, Factory Mutual Laboratories, Laboratory Report No. 11760, June 1950.

Croce, P. A.; "Modeling of Vented Enclosure Fires," VFDB 5th International Fire Protection Seminar, Karlsruhe, Germany, Vol. 1, pp. 23–37, Sept. 1976.

Culver, C., Kushner, J.; A Program for Survey of Fire Loads and Live Loads in Office Buildings (Tech. Note 858). National Bureau of Standards, Washington (1967).

Culver, C.; Survey Results for Fire Loads and Live Loads in Office Buildings, NBS-BSS 85, National Bureau of Standards, Washington, D.C. (1976).

Dardis, R.; "Cost Benefit Analysis of Flammability Standards;" AATCC National Technical Conference, p. 199–204, 1975.

Dean, R. K.; "Stored Plastics Test Program," *Journal of Fire and Flammability,* Supplement on Consumer Product Flammability, Vol. 2, pp. 248–313, Dec. 1975.

Denyes, W., and Quintiere, J.; "Experimental and Analytical Studies of Floor Covering Flammability with a Model Corridor;" J. Fire and Flammability/Consumer Product Flammability, *1*, 32–109, 1974.

Denyes, W., and Quintiere, J.; "Experimental and Analytical Studies of Floor Covering Flammability with a Model Corridor;" National Bureau of Standards NBSIR 73-199, 105 pp., May 1973.

de Ris, J.; "Modeling Techniques for Predicting Fires," Applied Polymer Symposium No. 22, J. Wiley, pp. 185–193, (1973).

de Ris, J., Kanury, A. M., and Yuen, M. C.; "Pressure Modeling of Fires." Fourteenth Symposium (International) on Combustion, p. 1033; The Combustion Institute, 1973.

de Ris, J., Kanury, A. M., and Yuen, M. L.; "Pressure Modeling of Fires;" Fourteenth Symposium (International) on Combustion, The Combustion Institute, pp. 1033–1044, (1973).

Draft copy of "Proposed Standard for the Flammability (Cigarette Ignition Resistance) of Upholstered Furniture" (PFF-6-76).

E-603-77 "Guide for Room Fire Experiments," The American Society for Testing and Materials. (1977).

Emmons, H.; "Heat Transfer in Fire;" Journal of Heat Transfer, pp. 145–151, May 1973.

Erling, Mark, and Reiser, J. W.; A Fire Service Demand Charge Study, NSF Grant No. ISR 74-191293-A; 2, City of Tacoma, Washington, Jan. 1976.

Fang, J. B.; Fire Buildup in a Room and the Role of Interior Finish Materials, NBS Technical Note 879, June 1975.

Federal Aviation Administration; "Compartment Interior Materials Toxic Gas Emission;" Federal Register, *39,* (251); p. 45044–45045, Dec. 30, 1974.

Federal Aviation Administration; "Smoke Emission From Compartment Interior Materials;" Federal Register, *40* (30), p. 6506–6509, Feb. 12, 1975.

Federal Fire Prevention and Control Act of 1974; Public Law 93—498, October 29, 1974.

Federal Test Method Std. No. 406, U.S. General Services Administration, Specifications Activity, Printed Materials Supply Division, Building 197, Naval Weapons Plant, Washington, D.C. 20407, Oct. 5, 1961.

Federal Trade Commission; "Cellular Plastic Products, Disclosure of Combustion Characteristics in Marketing and Certification;" Federal Register, Vol. 40, No. 142, p. 30842, July 23, 1975.

Fenimore, C. P., and Martin, F. J.; "Candle-type Test for Flammability of Polymers," Modern Plastics, *43,* p. 141 (1966).

Fire Protection Handbook, 14th edition, National Fire Protection Association, Boston, Mass. 1976.

Friedman, R., "A Survey of Knowledge About Idealized Fire Spread Over Surfaces;" Fire Research Abstracts and Reviews, *10,* pp. 1—8, (1968).

Friedman, R.; "Aerothermodynamics and Modeling Techniques for Prediction of Plastic Burning Rates;" J. of Fire and Flammability, *2,* pp. 240—256, (1971).

Friedman, R.; "Behavior of Fires in Compartments;" International Symposium on Fire Safety of Combustible Materials; University of Edinburgh, p. 100—113, October, 1975.

General Motors Corp., Fisher Body Div., Test Method 32—12 (August 16, 1971).

Hamilton, S. B.; "A Short History of the Structural Fire Protection of Buildings Particularly in England;" (Dept. of Scientific and Industrial Research, Special Report No. 27), HMSO, London (1959).

Hartzell, L. G.; "Development of a Radiant Panel Test for Flooring Materials;" J. of Fire and Flammability/Consumer Prod. Flammability *1,* pp. 376—389 (1974).

Hilado, C. J.; Flammability Test Method Handbook; Technomic Publishing Co.; Westport, Conn. 1973.

Hill, R. M.; "Operation School-Burning;" ibid, p. 296.

Holve, D. J., and Sawyer, R. F.; "Diffusion Controlled Combustion of Polymers;" Fifteenth Symposium (International) of Combustion, The Combustion Institute, pp. 351—361, (1975).

Huggett, C.; "Carpet Flammability and the NBS Corridor Fire Program;" Standardization News, *1* (5), p. 16—20, 1973.

Ingberg, S. H., Dunham, J. W., and Thompson, J. P.; Combustible Contents in Buildings, (Report BMS 149). National Bureau of Standards, Washington (1957).

Ingberg, S. H., *et al;* Fire Resistance Classifications of Building Constructions (Report BMS 92). National Bureau of Standards, Washington (1942).

Ingberg, S. H.; Q NFPA, 20, 243—252, (1927).

Ingberg, S. H.; Q NFPA, 20, 43—61, (1928).

Institute of Electrical and Electronics Engineers, Inc.; "IEEE Standard for Type Test of Class IE Electrical Cables, Field Splices, and Connections for Nuclear Power Generating Stations," IEEE Std. 383 (April 15, 1974).

Johnson, P. R.; "A General Correlation of the Flammability of Natural and Synthetic Polymers;" J. Applied Polymer Science, *18,* 491—504 (1974).

Lee, T. G., and Huggett, C.; "Interlaboratory Evaluation of ASTM E 84-70 Tunnel Test Applied to Floor Covering;" Journal of Testing and Evaluation, *3,* 3—14, 1975.

Lewis, B., and von Elbe, G.; "Combustion, Flames, and Explosions of Gases;" Academic Press, New York, 1961.

Lucha, G. V., Robertson, A. N., and Shooley, E. A.; "An Analysis of Aircraft Accidents Involving Fires;" NASA Final Report No. CR-13-7690, Stanford Research Institute, 1975.

Lyons, J. W.; "Requirements for Fire Safety of Combustible Materials in the United States;" International Symposium on Fire Safety of Combustible Materials; University of Edinburgh, October, 1975, p. 12—19.

REFERENCES

Magee, R. S., and McAlevy, R. F., III; "The Mechanism of Flame Spread;" J. Fire and Flammability, *2,* pp. 271–296, (1971).

Magee, R. S., and Reitz, R. D.; "Extinguishment of Radiation-Augmented Plastic Fires by Water Sprays;" Fifteenth Symposium (International) on Combustion, Combustion Institute, Pittsburgh, pp. 337–347, 1975.

Malhotra, H. L.; "An Examination of the Role and Philosophy of Fire Tests;" ISO, TC 92/WG 7, Doc. No. 58, July 1974.

Malhotra, H. L.; "The Philosophy and Design of Fire Tests;" International Symposium on Fire Safety of Combustible Materials; University of Edinburgh, October, 1975, p. 149–155.

Maroni, W. F.; "Rigid Cellular Wall Insulations;" Fire Journal, Nov. 1972, p. 38.

Martin, F. J.; "A Model of Candle-like Burning of Polymers;" Combustion and Flame *12,* 125–135 (1968).

Kanury, A. M.; "Theoretical Analysis of Fire and Flammability Tests – The Limiting Oxygen Index Test;" International Symposium on Fire Safety of Combustible Materials, Edinburgh, p. 187–198, 1975.

Kashiwagi, T.; "A Study of Flame Spread Over a Porous Material Under External Radiation Fluxes;" Fifteenth Symposium (International) on Combustion, The Combustion Institute, pp. 255–265, (1975).

Kim, J. S., de Ris, J., and Krœsser, F. W.; "Laminar Free-Convective Burning of Fuel Surfaces;" Thirteenth Symposium (International) on Combustion, Combustion Institute, pp. 949–961, (1971).

Kline, L. M.; "Burning, Arcing, Ignition, and Tracking of Plastics Used in Electrical Appliance;" Underwriters Laboratories Research Bulletin No. 55 (February 1964).

Kung, H. C.; "Residential Sprinkler Protection Study;" Factory Mutual Research Corporation Serial 22442, November 1975.

N.A.S.-N.R.C. Publication 786, "The Use of Models in Fire Research," Washington, D.C. 1961.

The National Building Code, 1976 edition, American Insurance Association, Engineering and Safety Service, 85 John St., New York, N.Y. 10038.

National Bureau of Standards; "Consideration in the Use of Sampling Plans for Effective Compliance with Mandatory Safety Standards;" NBSIR 75-697, June 1975.

N.B.S. Special Publication 357, "The Mechanisms of Pyrolysis, Oxidation, and Burning of Organic Materials;" G.P.O., Washington, D.C., 1972.

National Bureau of Standards; "Directory of United States Standardization Activities;" NBS Special Publication 417, 1975.

National Bureau of Standards Special Publication 417, Directory of U.S. Standardization Activities, U.S. Government Printing Office, p. 2, 1975.

National Bureau of Standards; "Report of the Voluntary Standardization Policy Study Group;" NBS Report 10 391, December, 1970.

National Fire Prevention and Control Administration; "Highlights of the National Household Fire Survey," 1975.

Nat. Fire Protection Assoc., NFPA 701, Nat. Fire Codes, Vol. 3, 701-1-24 (1967–1968).

National Fire Protection Association; "Fires and Fire Losses Classified, 1974;" Fire Journal, *69* (5), p. 43, 1975.

National Fire Protection Association, National Fire Codes, Vol. 15, 258-1 (1976).

National Materials Advisory Board, Materials and Processes, Specifications and Standards, NMAB-330, National Research Council, Washington, D.C. 1977. (Available from National Technical Information Service, Springfield, Va. 22151.

Nelson, G. L., and Webb, J. L.; "Oxygen Index of Liquids: Technique and Applications;" J. Fire and Flammability *4,* 210–226 (1973).

Orloff, L., de Ris, J., and Markstein, G. H.; "Upward Turbulent Fire Spread and Burning of Fuel Surfaces;" Fifteenth Symposium (International) on Combustion, The Combustion Institute, pp. 183–192, (1975).

Orloff, L., Modak, A. T., and Alpert, R. L.; Burning of Large-Scale Verticle Surfaces;" Sixteenth Symposium (International) on Combustion, p. 1345, The Combustion Institute, 1976.

Parker, W. J., and Long, M. E.; Ignition, Heat Release, and Non-combustibility of Materials, ASTM ST 502, American Society for Testing and Materials, pp. 135–151, 1972.

Quintiere, J. G.; "A Characterization and Analysis of NBS Corridor and Performance of Floor Covering Materials;" National Bureau of Standards NBSIR 75-69, (June 1975a) 83 pp.

Quintiere, J. G.; "Some Observations of Building Corridor Fires;" Fifteenth Symposium (International) on Combustion, The Combustion Institute, Pittsburgh, Pa., pp. 163–174, (1975b).

Reimschuessel, H. K., Shalaby, S. W., and Pearce, E. M.; "On the Oxygen Index of Nylon 6;" J. Fire and Flammability *4,* 299–308 (1973).

Roberts, A. F.; "Extinction Phenomena in Liquids;" Fifteenth Symposium (International) on Combustion, p. 305–313 (1975).

Robertson, A. F., Gross, D., and Loftus, J.; A Method for Measuring Surface Flammability of Materials Using a Radiant Energy Source, Proc. Am. Soc. Testing and Materials, *56,* 1437–1453 (1956).

Robertson, A. F.; "Test Method Categorization and Fire Hazard Standards;" ASTM Standardization News, *3* (11), p. 18–20, 1975.

Rodak, S., and Ingberg, S. H.; Full-Scale Residential Occupancy Fire Tests of 1939. Unpublished NBS Report (1976).

Routley, A. F.; "The Development of the Oxygen Index Concept for the Assessment of the Flammability Characteristics of Materials;:: Central Dockyard Laboratory, HM Naval Base, Portsmouth, Report No. 573 (1973).

Shorter, G. W.; "The St. Lawrence Burns;" The Use of Models in Fire Research, p. 289; National Academy of Sciences – National Research Council, Washington, D.C., 1961.

Simmons, R. F., and Wolfhard, H. G.; "Some Limiting Oxygen Concentrations for Diffusion Flames in Air Diluted with Nitrogen;" Combustion and Flame, *1,* 155–161 (1957).

Smith, E. E.; Heat Release Rate of Building Materials, ibid, pp. 119–134.

Society of Automotive Engineers, Tech. Rept. J369 (March 1969).

Spalding, D. B.; "The Art of Partial Modeling;" Ninth Symposium (International) on Combustion; Academic, New York, p. 833–843, 1963.

Standard Specifications for Fire Tests of Materials and Construction (C 19-18). American Society for Testing and Materials, Philadelphia (1918).

Steiner, A. J.; Method of Fire Hazard Classification of Building Materials, ASTM Bulletin No. 121, 19–21 (1943).

Steiner, A. J.; Symposium on Fire Test Methods; p. 80; ASTM Special Technical Publication No. 301, 1961.

State of California, Department of Consumer Affairs, Bureau of Home Furnishings, Technical Information Bulletin No. 117, February 1975, Requirements, test procedure and apparatus for testing the flame retardance of upholstered furniture. Effective 1, 1975.

Stuetz, D. E., Barnes, B. P., DiEdwardo, A., and Zitomer, F.; "Flammability as an Intrincis Property of Polymers;" Presented at the 42nd TRI Annual Research and Technology Conference, New York City, April 5–7, 1972.

Tewarson, A.; "Experimental Fires in Enclosures: I. Cellulosic Materials;" Combustion and Flame, *19,* pp. 101–111, (1972).

Tewarson, A.; "Flammability of Polymers and Organic Liquids – Part I. Burning Intensity;" Factory Mutual Research Corporation Report RC 75-T-6, February 1975.

Textile Flammability, American Association of Textile Chemists and Colorists, 1975.

Thompson, Norman J., and Cousins, E. W.; The FM Construction Materials Calorimeter, NFPA Quarterly, *52,* No. 3, 186–96 (Jan. 1959).

REFERENCES

Tu, K. M., and Davis, S.; "Flame Spread of Carpet Systems Involved in Room Fires;" NBSIR 76-1013, 1976.

Underwriters Laboratories, "Rubber-Insulated Wires and Cables, Std. UL 44 (July 11, 1968).

Underwriters Laboratories, "Tests for Flammability of Plastic Materials for Parts in Devices and Appliances;" Std. UL 94 (February 1, 1974).

Underwriters Laboratories, "Thermoplastic-Insulated Wires;" STD. UL 83 (December 24, 1975).

U.S. Consumer Products Safety Commission, Policy on Voluntary Standards, News Release, November 10, 1977.

U.S.D.A. Forest Products Laboratory Report 2257 "Surface Flammability as Determined by the FPL 8-foot Tunnel Method," 1962.

U.S. Dept. of Commerce, "Flammability Standard for Mattresses;" DOC FF 4-72, Federal Register, Vol. 38, No. 110, 15095–15101 (June 8, 1973).

U.S. Dept. of Commerce, "Standards for Surface Flammability of Carpets and Rugs;" DOC FF 1-70, Federal Register, Vol. 35, No. 74, 6211–6214 (April 16, 1970).

U.S. Federal Highway Admin., Motor Vehicle Safety Standard No. 302, Federal Register, Vol. 36, No. 5, 289–291 (January 8, 1971).

U.S. Gen. Services Admin., "Floor Coverings, Resilient, Nontextile;" Fed. Test Method Std. No. 501 (June 15, 1966).

U.S. Gen. Services Admin., Federal Test Method Std. No. 406 (October 5, 1961).

Vickers, A. K.; "The Flammable Fabrics Accident Case and Testing;" in Materials Information Program: An Interagency Review of Federal Agency Activities on Technical Information About Materials: Paper VII-2, NBS SP463, 1977.

Waterman, T. E.; "Room Flashover-Criteria and Synthesis;" Fire Technology, *4,* p. 25–31, 1968.

Williams, F. A.; "Combustion Theory;" Addison-Wesley, Reading, Massachusetts, 1965.

Williamson, R. B., and Baron, F. M.; "A Corner Fire Test to Simulate Residential Fires;" J. of Fire and Flammability, Vol. 4, p. 99, April 1973.

Williamson, R. B.; Department of Civil Engineering, Unversity of California, Berkeley, Private Communication 1976.

Wilson, W. J.; "Large Scale Fire Tests;" J. Fire and Flammability, *7,* p. 112–124, 1976.